T0206263

Lecture Notes in Physics

Founding Editors

Wolf Beiglböck

Jürgen Ehlers

Klaus Hepp

Hans-Arwed Weidenmüller

Volume 1001

Series Editors

Roberta Citro, Salerno, Italy

Peter Hänggi, Augsburg, Germany

Morten Hjorth-Jensen, Oslo, Norway

Maciej Lewenstein, Barcelona, Spain

Angel Rubio, Hamburg, Germany

Wolfgang Schleich, Ulm, Germany

Stefan Theisen, Potsdam, Germany

James D. Wells, Ann Arbor, MI, USA

Gary P. Zank, Huntsville, AL, USA

The series Lecture Notes in Physics (LNP), founded in 1969, reports new developments in physics research and teaching - quickly and informally, but with a high quality and the explicit aim to summarize and communicate current knowledge in an accessible way. Books published in this series are conceived as bridging material between advanced graduate textbooks and the forefront of research and to serve three purposes:

- to be a compact and modern up-to-date source of reference on a well-defined topic;
- to serve as an accessible introduction to the field to postgraduate students and non-specialist researchers from related areas;
- to be a source of advanced teaching material for specialized seminars, courses and schools.

Both monographs and multi-author volumes will be considered for publication. Edited volumes should however consist of a very limited number of contributions only. Proceedings will not be considered for LNP.

Volumes published in LNP are disseminated both in print and in electronic formats, the electronic archive being available at springerlink.com. The series content is indexed, abstracted and referenced by many abstracting and information services, bibliographic networks, subscription agencies, library networks, and consortia.

Proposals should be sent to a member of the Editorial Board, or directly to the responsible editor at Springer:

Dr Lisa Scalone
lisa.scalone@springernature.com

Kazunori Hanagaki · Junichi Tanaka ·
Makoto Tomoto · Yuji Yamazaki

Experimental Techniques in Modern High-Energy Physics

A Beginner's Guide

Springer

Kazunori Hanagaki
Institute of Particle and Nuclear Studies
KEK
Ibaraki, Japan

Makoto Tomoto
Institute of Particle and Nuclear Studies
KEK
Ibaraki, Japan

Department of Physics
Nagoya University
Nagoya, Japan

Junichi Tanaka
International Center for Elementary Particle
Physics
The University of Tokyo
Tokyo, Japan

Yuji Yamazaki
Department of Physics
Kobe University
Kobe, Japan

This work was supported by Sponsoring Consortium for Open Access Publishing in Particle Physics.

ISSN 0075-8450   ISSN 1616-6361   (electronic)
Lecture Notes in Physics
ISBN 978-4-431-56929-9   ISBN 978-4-431-56931-2   (eBook)
https://doi.org/10.1007/978-4-431-56931-2

This Springer imprint is published by the registered company Springer Japan KK, part of Springer Nature.
The registered company address is: Shiroyama Trust Tower, 4-3-1 Toranomon, Minato-ku, Tokyo 105-6005, Japan

Preface

This book is meant to be a practical introduction to data analysis in high-energy physics experiments, especially collider experiments using high-energy accelerators.

The field requires a very wide range of knowledge, not only for the theoretical particle physics but also for the detector technology and computing science. We often find beginners in this field suffering from understanding how data analysis is taking place, simply because of too many things one needs to know. Reading journal papers often does not help since comprehensive understanding and training are required, which are not described in the papers themselves. It is quite difficult to obtain such skills unless you do once a data analysis by yourself. We hope that this book helps reduce such difficulties by providing a one-stop "explanation" on key aspects of data analysis.

This book should also serve as an introductory textbook for those who are learning about individual subjects in data analysis, such as

- Basic idea on methods to reconstruct and identify particles;
- Detector calibration in collider experiments;
- Statistical methods used in collider experiments;
- Methods to increase sensitivities of an experiment through data analysis techniques;
- Simulation of particle collisions and detector responses.

This book is intended for undergraduate and first-year graduate students who have taken basic-level courses in particle physics. Throughout the book, we tried to explain what happens without explicitly using many equations. We neither cover the formalism on the interaction of particles in matter nor the theory of the particle physics Standard Model. Instead, we, experimental physicists, provide a "practical" explanation of how to understand after considering those formalism and theories.

Tsukuba, Japan   Kazunori Hanagaki
Tokyo, Japan   Junichi Tanaka
Nagoya/Tsukuba, Japan   Makoto Tomoto
Kobe, Japan   Yuji Yamazaki
October 2021

Acknowledgments

We thank Springer for giving us a chance to publish this book. We greatly appreciate Springer's editors for their patience. The start of writing this book is back in 2015 when particle physicists basked in the afterglow of the discovery of the Higgs boson. Since then our writing was so slow, but the editors kept us encouraging to go on. Without their endless efforts, this book would not be realised.

We thank our families for their continuous support not only in writing this book but also the physicists' life.

Tsukuba, Japan   Kazunori Hanagaki
Tokyo, Japan   Junichi Tanaka
Nagoya/Tsukuba, Japan   Makoto Tomoto
Kobe, Japan   Yuji Yamazaki
October 2021

Contents

About the Authors

Kazunori Hanagaki has been a Professor at KEK since 2015. He received his master's (1995) and Ph.D. (1998) degrees from Osaka University for the study of CP violation in Fermilab KTeV experiment. He worked on the Belle experiment as a postdoc researcher at Princeton University and the Dzero experiment as Wilson Fellow and Scientist at Fermilab and joined ATLAS experiment when he became an Associate Professor ar Osaka University. His current research interest is in Higgs physics and detector development. He is one of the ATLAS Japan group co-spokespersons since 2015.

Junichi Tanaka has been a Professor at International Center of Elementary Particle Physics (ICEPP), the University of Tokyo (UTokyo) since 2018. He received his master's (1999) and Ph.D. (2002) degrees from UTokyo for the studies in the Belle experiment. In 2002, he joined the ATLAS collaboration to search for Higgs boson and new physics beyond the standard model. He was involved in the Higgs search with $H \to \gamma\gamma$ and largely contributed to the discovery of the Higgs boson in 2012. Then, he worked on the ATLAS calorimeter upgrade, SUSY search, etc. with his master's and Ph.D. students and his colleagues. He also has worked on computing for high-energy physics: grid, cloud, artificial intelligence, quantum computers, etc. He also joined the Particle Data Group as an encoder of the standard model Higgs since 2017.

Makoto Tomoto has been a Professor at KEK since 2020. He received his master's (1994) and Ph.D. (2001) degrees from Nagoya University for the studies of the track trigger with central drift chamber and $B_d^0 \bar{B}_d^0$ mixing at the Belle experiment. He worked on Tevatron/Dzero experiment as a Research Associate at Fermilab from 2001 to 2006 and joined the ATLAS experiment as an Associate Professor at Nagoya University in 2006. He devoted his large efforts to the measurements of the top-quark pair production cross-section at the beginning of the ATLAS experiment. His current research is focused on the Higgs physics, top-quark physics, and the upgrade of the first-level muon trigger system. He is also involved in detector developments for future collider experiments.

Yuji Yamazaki has been a Professor at Kobe University since 2014. He joined the ATLAS experiment in 2007 when he moved to Kobe and has been engaged in software-based high-level trigger system in ATLAS for muons and recently in muon detector upgrade for the high luminosity LHC. He has also been working on Higgs and top-quark production cross-section measurements. Before starting the LHC experiment, he was involved in the HERA experiment, a lepton-proton collider at the DESY laboratory in Germany, when he was a Research Associate at KEK. There he worked on the physics of deep-inelastic scattering, especially on hadronic final state (jets, etc.) and diffractive scattering between the virtual photon and the proton.

Introduction

Experimental techniques in high-energy particle physics have been developing very rapidly, despite the experimental principle being relatively simple. Here, we explain the principle first, then the development on top, since the experiments are more complex as the energy of collisions increases.

The purpose of the high-energy experiments can be classified roughly into two categories: to find any new particle by converting the collision energy to particle mass, and to investigate the nature of the interaction in order to find if any new feature exists there. The first category produces a new particle through *resonance* or radiation in the final state. The second category may be an indirect detection of new particle contribution in the interaction or discovery of sub-structure of the "elementary" particles through precise measurements of, for example, scattering angles.

Figure 1.1 shows two patterns (Feynman diagrams) of e^+e^- collisions corresponding to the two categories of experiments. The most typical interaction of the first category experiments, production of a new particle, is the resonance of state X in e^+e^- annihilation as depicted in Fig. 1.1a. In most or all collisions, the state X corresponds to known particle(s), for example $Z^{(*)}/\gamma$, but may contain a contribution from a new particle. If there is a contribution from such a new particle, the invariant mass of the decay product reconstructed using the energy and momentum of all the decay particles from X may show a peak from the resonance, corresponding to the mass of the unknown. As the simplest example, if we assume that X decays into two particles, we can observe a peak in the invariant mass spectrum of the two particles.

For the second category of the experiments, namely to investigate interactions and sub-structure of particles, we instead like to use processes where an incoming particle scatters off the other via an exchange of a state X' as represented by Fig. 1.1b. Depending on the nature of the exchange X', the scatters will give a certain prediction on the angular and momentum distribution of the final state particles. For example, for

© The Author(s) 2022
K. Hanagaki et al., *Experimental Techniques in Modern High-Energy Physics*,
Lecture Notes in Physics 1001, https://doi.org/10.1007/978-4-431-56931-2_1

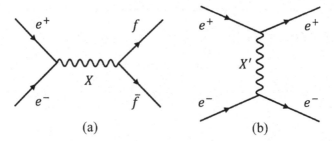

Fig. 1.1 Feynman diagrams of e^+e^- interactions **a** annihilating each other producing a virtual particle X (decaying a pair of particles) and **b** exchanging a particle X' between them

the Coulomb scattering, we know that the scattering angle of an electron beam with a heavy point-like target particle, such as nucleus, behaves like $1/\sin^4(\theta/2)$, where θ is the scattering angle with respect to the incoming beam direction in target-rest frame (note that the spin effect to the scattering angle is ignored here). The contribution from a new particle exchanged as the state X' would give a small modification to the fundamental Rutherford scattering behaviour of $1/\sin^4(\theta/2)$. The effect should be enhanced beyond the energy regime where the momentum transfer squared of the exchanged particle is beyond the mass squared of the particle responsible for the new interaction, such as an unknown new heavy gauge boson.

We conduct high-energy experiments for research based on these principals and need to consider detector, data-taking, reconstruction, identification, calibration, analysis, etc. The $2 \rightarrow 2$ processes, as illustrated in Fig. 1.1a, b, can be fully reconstructed if the momenta and energies of these two outgoing particles are measured; the experiment for such a case is simple. In reality, the number of particles to be measured may be more than two in many cases—in fact, far more for typical high-energy scatterings. There may be the radiation of particles from the 2-to-2 process as well as decay products of heavy elementary particles, such as W and Z bosons and top quarks, which decay further into many particles.

The situation is particularly difficult if particles from the outgoing particles may contain neutrinos or any other unknown neutral particles which interact only weakly with detector materials. Such neutral particles escape the detector, practically always. The only way to "detect" them is to measure the so-called "missing momentum" using four-momentum conservation, which corresponds to momentum carried by the neutral particles. For that, we need to measure all the other particles in the final state. The detector, therefore, needs to cover the interaction point almost fully, i.e. the solid angle of the coverage should be close to 4π. Such kind of detector is called a hermetic detector. A neutrino becomes a common object once the energy of the collision is high enough that one can easily produce W and Z bosons, since they produce neutrinos through the decay $W \rightarrow \ell\nu$ or $Z \rightarrow \nu\bar{\nu}$ where ℓ is either of e, μ, τ and $\nu = \nu_e, \nu_\mu$ or ν_τ. This is the reason why any modern high-energy collider experiments cover almost all the solid angles to have a hermetic system.

The density of particles in the detector is also an issue at high energies. As the energy of the collision becomes higher, the number of particles increases approximately proportional to $\ln \sqrt{s}$, where s is the square of the centre-of-mass energy of the collision. This makes the angular density higher; in particular, many particles are produced in a small angular area when an energetic quark or gluon is produced and fragmented. These collimated bunches of particles are called *jet*. The presence of jets also requires the detector to have small segmentation—or fine "granularity", we often call—such that a pair of particles produced close by each other can be distinguished as two particles. In addition, the increase of the collision energy also asks for more material to stop neutral particles (not neutrinos but neutrons) in order to measure them. The detector becomes thicker with energy, again $\propto \ln \sqrt{s}$, and the overall size of the detector becomes larger.

Furthermore, the modern high-energy experiments should deal with many different types of stable particles. The detector has to measure electrons, muons and photons very precisely. Hadrons (pions and kaons, practically) are also copiously produced as explained above. The energy measurement of charged hadrons or baryons at high precision is particularly difficult. The identification of the species of hadrons, such as the pion, the kaon and other particles, may be desirable if one needs to study decay chains of particular mesons.

In addition, the presence of b-quark is a very useful signal for investigating physics involving quark flavour, in particular for top quarks, since a top quark decays to W and a b-quark, whose branching ratio is practically 100%. As for leptons, production of high momentum τ could be an indication of the presence of new physics of special flavour structure, such as some bosons preferentially couple to third generation fermions. Identification of these particles is based on the fact that they fly short distance before decaying, calling for a very precise tracking device.

Thus, modern high-energy collider detectors should be able to deal with all sorts of stable leptons and hadrons. In general, the design of detectors should be optimised to the target physics. For collider detectors, however, the versatility is more respected since there are many different targets; detectors need to measure known Standard-Model (SM) processes precisely while they also cope with some peculiar signals from new particles. Therefore, "general-purpose" detectors are preferred and measurements on both known and unknown processes are performed with good precision.

Yet another to consider at high energy is that the probability of observing particular processes, i.e. cross sections (see Sect. 2.2) decreases with energy of collisions in general. A simple dimensional analysis tells us that the cross section of point-like particles, such as e^+e^- or parton-parton collisions, decreases like $1/s$, or by $1/E^2$ in terms of the incoming particle energy E for symmetric collisions. This means that the probability of interactions you like to find is suppressed by $1/E^2$. This should be compensated by increasing luminosity (see Sect. 2.2), which is proportional to the number of collisions of a given process per unit time. This is realised by a shorter time interval of collisions, higher beam current and better focusing of beams in colliders.

Now the problem is that this increases not only signal but also background rates. In particular, for hadron-hadron collisions, the cross section is dominated by a soft process (see Sect. 2.5), which is approximately constant in collision energy instead of $1/E^2$. The soft process is the background for most physics analyses. The increase in the number of collisions per unit of time also leads to a higher probability of pile-up of multiple collisions. The issue of pile-up should be resolved by a detector system with high resolution, both in space and time. Moreover, higher rates impose an additional challenge for data-taking.

Last but not the least, for precision measurements it is also important that experiments are well modelled by simulation. So-called data-driven methods are adopted to reduce uncertainties of measurements for the estimation of background contributions. However, in most cases, we need help with the simulation, which causes additional uncertainties. In data analysis, the cutting-edge technique utilising the modern statistical technique including machine learning should help in increasing the experimental sensitivities.

In summary, modern high-energy experimental physics, specifically the collider experiments, should pay attention to the following items, even though the basic idea of experiments in terms of physics goal remains similar to that of lower energy experiments. Good reconstruction of all sorts of stable particles including neutrinos should be possible with well-calibrated detectors. The detector should also work under harsh conditions of high-rate collisions. Modern statistical and analysis methods and well-modelled simulation of physical processes and detectors should be pursued with support from the rapid advancement of computing powers. Last but not the least, the sensitivity of the experiments ultimately relies on ideas on data analysis based on the human understanding of the physics processes in concern. We believe it even if smarter artificial intelligence is born.

The following chapters of this book provide comprehensive explanations on each of the key elements in the high-energy experiments and data analysis, aiming for helping to understand on the above-mentioned subjects. Chapter 2 starts with an overview of how a collider detector is designed to measure particles and the data are recorded. Also explained is how we extract the physical quantities of interest out of the data analyses. Chapter 3 covers the collider facility, detector in general and data-taking system. Chapter 4 gives basic overview of statistics used in high-energy physics. Chapter 5 describes detector calibration procedure, followed by particle identification in Chap. 6. Chapter 7 is devoted to the explanation of how the simulation of the physical process of collisions is taken place. All these chapters are followed by "exercise" parts in Chap. 8, where we give an explanation of the physics data analysis and results of journal papers as examples on how the reconstructed events are utilised to extract physical properties. They are measurements of Higgs production cross sections through its decay to two photons, a $b\bar{b}$, or a W^+W^- pair. Searches for new particles are also explained.

Basic Idea of Measurements in Particle Collisions

2

2.1 Observables of Particle Scattering

Some of the readers of this book may have seen a so-called "event display", visualisation of a particle collision. An example from the ATLAS experiment is given in Fig. 2.1. There, what we see are many curves from a particular point, which indicates the location of two particles colliding with each other. The curves emerging from the point are charged tracks, and the trace of charged particles are identified from detector responses. Also seen are many colourful boxes, which look like histograms, with the direction of the height of the histograms pointing to a radial direction. These indicate the amount of energy from particles produced via the collision, measured in particle detectors.

We like to extract the properties of the physics processes of collision through such measurements. Ultimately, we like to know the property of the underlying interactions, i.e., how the particles are scattered and the final state particles are produced at the level of a few elementary particles involved. As you know, however, we have no way to know exactly how the elementary particles are interacting since it is through a process obeying quantum mechanics. All we can do experimentally is to measure the physical observables as precisely as possible in order to obtain the distribution of the observables. Such observables include the species of the particles produced by the collisions and the energy, momentum, and possibly the quantum numbers of the particles. Then we identify the underlying interaction of particles by comparing theoretical predictions of various interactions.

Here, we explain how such a correspondence between the observable and the underlying process is realised by taking again an example of the reaction $e^+e^- \rightarrow e^+e^-$ (Fig. 2.2), in e^+e^- colliders. In the experiment, we can measure the existence of the final state particles, e^+ and e^-, as well as their energy and momentum.

Let us suppose that you have no knowledge of quantum electrodynamics (QED). Then you do not know how the electrons exchange force when they scatter, i.e.

© The Author(s) 2022
K. Hanagaki et al., *Experimental Techniques in Modern High-Energy Physics*,
Lecture Notes in Physics 1001, https://doi.org/10.1007/978-4-431-56931-2_2

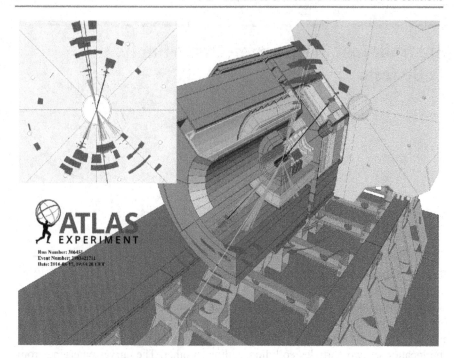

Fig. 2.1 Event display for an event, where the invariant mass of the two electrons found in the event is one of the highest among the events (up to 2016) taken by the ATLAS experiment. Reprinted under the Creative Commons Attribution 4.0 International license from [1] © 2017 CERN for the benefit of the ATLAS Collaboration. The black lines (or curves, more precisely) represent the charged tracks for the two electron candidates. The yellow curves correspond to other charged tracks. The red and purple boxes indicate energies measured in calorimeter units

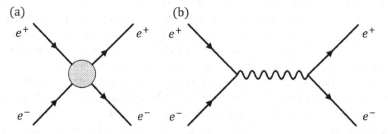

Fig. 2.2 Diagram for a reaction $e^+e^- \rightarrow e^+e^-$. **a** the detail of the interaction is left unspecified. **b** An example of the lowest order Feynman diagram

how the force is mediated. We often draw interactions of particles schematically in diagrams. The diagram is drawn with a blob at the vertex of four fermions, in this case either electron or positrons. The interaction of the objects inside the blob is not "visible": although we can guess that some particles are exchanged to mediate the scattering force, they cannot be observed directly. All we can observe are the "observables", the particles in the final state.

Now we like to know what is happening in the blob through experimental measurements of the scattering. If we observe just one event of the interaction $e^+e^- \to e^+e^-$, there is not much more information than that such a kind of interaction exists. However, if we observe many events with an e^+e^- pair in the final state, we can also measure the distributions of the final state particles. This would give much more information on the interaction. The distribution can be angular, energy or, more generally, the (four-)momentum distributions. Not only the shape of the distribution but also the integrated number of events for a given intensity of collisions (the luminosity, see the next section) tell us the "strength" of the interaction, i.e. how often such interactions should occur. For simple interactions like $e^+e^- \to e^+e^-$, the angular and momentum distributions allow us to extract the shape of the potential between two scattering particles through appropriate transformation. For more complicated interactions like collisions with many particles produced, inclusive or exclusive distributions (see Sect. 2.3) of a particular type of particles can be compared to theoretical predictions. In this way, the blob of the diagram is uncovered and the Lagrangian of the interaction could be constructed, even though the interpretation is often limited by both experimental uncertainties from the measurement and uncertainties of the theoretical predictions.

2.2   Cross Section and Luminosity

The strength of the scattering has the dimension of L^2 (area), where L represents the dimension in length. This can be understood if we take an example of scattering in classical mechanics with one of the objects standing still (a target). Suppose that you hit the target of a certain size by throwing an object (a projectile) moving in a certain direction (Fig. 2.3a). If you assume that the object can hit only when they contact each other, and assume that the size of the projectile is negligible, the probability to hit the target is proportional to the size viewed from the direction where the projectile is running. More precisely, it is proportional to the cross section σ of the object on the plane perpendicular to the projectile direction. For that reason, we also call the size of the interaction area the cross section. Whether the projectile goes inside the interaction area or not depends on the distance between the centre of the target and the line where the projectile flies. The distance between the target centre and the trajectory of the projectile at the infinite distance projected to the area of the target in the plane perpendicular to the projectile momentum is called the impact parameter b (see Fig. 2.3b).

For point masses that interact remotely, which is the case for the elementary particles, the projectile and target are scattered infinitesimally even if their impact parameter is very large, if the force is long-distance, i.e. propagating in the infinite distance. This implicates that the cross sections for such an interaction are well defined as a function of a property of the scattering, such as the scattering angle θ in Fig. 2.3. Differential cross sections can be defined using the variable: $d\sigma/d\theta$ in this case.

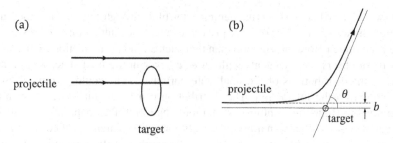

Fig. 2.3 a Target and projectile. **b** Scattering angle θ and impact parameter b, where a projectile particle is injected towards a target particle

The differential cross-section formula can be calculated using theory if the interaction is known. If we observe any deviation in such a distribution from the prediction, this tells us that the particle is not point-like or there exist some new type of interaction.

The number of observed scatterings for a given cross section can be calculated if we know the "strength" of the projectile beam. One needs to give as many particles in an area of as small cross section as possible, in order to maximise the number of events. This means that the particle flow density $(1/a)dN_p/dt$ is one of the parameters to define the number of interactions. Here, a is the area where the projectile is injected, i.e. the size of the beam of the projectile. N_p gives the number of projectiles passing through the area. Then the number of interactions N per unit time is given as $dN/dt = \sigma \cdot (1/a)dN_p/dt$. You see that the cross section gives the correct dimension to deduce the number of interactions.

To obtain the number of collisions, we need to also express the number of particles in a target object via the density of the target. For the colliding beam, the corresponding number would be the density of the "target beam". For the fixed target, one should count the number of target particles within the projectile beam size a. Suppose that the depth of the target is D and the target number density per volume is n_t. The total number of targets N_t, which could be scattered by the projectile beam, is $N_t = Dan_t$. It is useful to express the target density in terms of the mass density of material ρ when we discuss the interaction of a particle with material such as that consisting of detectors. The number of the target is then expressed as

$$N_t = Da\frac{\rho N_A}{A},$$

where the targets are nucleus and ρ given is in g/cm^3, A is the mass number given in g/mol, and N_A is the Avogadro number. The number of the interactions for a still target (for the case of the fixed target experiment) integrated over time is reduced to

$$N = \frac{N_p N_t \sigma}{a} = \frac{N_p D \sigma \rho N_A}{A},$$

assuming the target is larger than the beam size.

The number of collisions for the collider can be obtained by replacing the number of targets with the number of the beam particle. For the beam as a target, it is easier to use the number of target beam particles per unit time dN_t/dt since the target is also not a still object. The beam particles are bunched in RF buckets (see Sect. 3.2) for almost all colliders. The number of particles per unit time crossing a given plane perpendicular to the beam dN_{beam}/dt is given as $f_{\mathrm{coll}} \cdot n_{\mathrm{bunch}}$ where f_{coll} is the number of bunch collisions per unit time and n_{bunch} is the number of particles in a bunch. In an ideal situation where both the lateral bunch size a (dimension= L^2) and the longitudinal distribution of particles inside the bunch are the same for both the target beam and the projectile beam, the frequency of collisions is given as

$$\frac{dN}{dt} = \sigma \cdot f_{\mathrm{coll}} \frac{n_1 n_2}{a},$$

where n_1 and n_2 are the numbers of particles in a bunch for beam 1 and beam 2—we no longer distinguish the target and projectile beams at this point. This implies that it is convenient to define the instantaneous luminosity L,

$$L \equiv f_{\mathrm{coll}} \frac{n_1 n_2}{a}.$$

Note that the dimension of L is $L^{-2}T^{-1}$. Then the number of collisions N_{coll} for a given period with the cross section σ can be obtained as

$$N_{\mathrm{coll}} = \sigma \times \int L dt,$$

where $\int L dt$ is called integrated luminosity.

2.3   Identifying Processes Through Measurements of Final State Particles

In this section, we discuss the way to identify if the observed events are indeed the ones of concern. You may regard the identification of events is just as simple as counting the number of events with a given final state, if we can safely assume that the detector works well enough to identify the particles. There are, however, a few points yet when we consider the "definition of the signal events."

Let us start with the example of $e^+e^- \rightarrow e^+e^-$. To see if an event is classified to this category or not, first we need to identify the type of the final state particles through the measurements by detectors. We call this procedure particle identification or *particle ID*. We discuss the technical detail of the particle ID in Chap. 6. Here, we simply assume that the final state particles are identified at certain probabilities. You would then count the number of particles of interest. In this example, you would request that the events should have one electron and one positron in the final state.

A few questions come along: we may wonder if we should request certain criteria in energy or momentum for the electron and positron. Also, we need to decide if we allow any other particle(s) in the final state. For example, since what we like to measure is the $e^+e^- \to e^+e^-$, you may like to limit yourself to select the events which have only one pair of e^+e^- and no additional particles are observed. In order to have such a selection well-defined, we need to consider the following things.

Firstly, the electrons can emit soft photons induced by an internal electromagnetic field induced by the interacting electron and positron themselves. This probability is very high for the electrons, whose mass is very small. Since such soft photons are anyhow not efficiently observed by the detector, experimentally you can define that the event contains only an e^+e^- pair *observed* in the final state by requesting the other detector responses not associated with either of the electron or positron to be below certain thresholds (typically slightly above the noise level of the detectors). A photon collinearly emitted in the direction of the electron would still be difficult to separate from the parent electron. In that case, such collinear photons are often treated as a part of the parent electron and then the measured energy and momentum are considered to be those of the primary electron.

The second reason is that none of the particle detectors using accelerators have 4π coverage in the solid angle: the detector is not completely hermetic. This means that we may miss a part of final state particles even if it is hard enough to be detected since we may have holes in the detector in that direction. For example, the probability for an e^+e^- collision to emit a photon in the direction of the incoming electron or positron is very high, again because of the small mass of the electron. Such events are called initial state radiation (ISR) events. It is not possible to catch the photon if the emission is at a very small angle from the incoming beam direction since the accelerator should accommodate the beam with a beam pipe with the finite diameter of typically more than a few centimetres. The photon escapes from the detector through the beam pipe.

Therefore, all we can do to select an event is to impose criteria which look like an e^+e^- final state. For an e^+e^- collider where the laboratory frame coincides with the centre-of-mass system of the two beams, which have the same energy, an example of such criteria would be

- a pair of electron-like particles in opposite charge, both within the angular region $\theta_{min} + \Delta < \theta < \pi - (\theta_{min} + \Delta)$. Here, θ is the polar angle with the z direction defined as the incoming direction of one of the beams, θ min, which is the polar angle of the detector boundary of the hole to accommodate the beam pipe, and Δ is a margin to be taken so that the observed particles are enough away from the boundary;
- no other track nor cluster observed in other parts of the detector with their momentum or energy greater than $0.05E_{beam}$, where E_{beam} is the beam energy; and
- each electron fulfils $E_{elec} > 0.8E_{beam}$, where E_{elec} is the energy of the electron or the positron.

The first criterion makes sure that the events are well contained within the angular coverage of the detector $\theta_{min} < \theta < \pi - \theta_{min}$. The second criterion removes the events where extra particles with significant energies on top of the signal electron

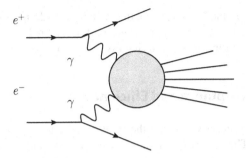

Fig. 2.4 The two-photon process in e^+e^- collisions. The blob indicates an interaction of incoming $\gamma\gamma$, emitted from e^+ or e^-, and outgoing final-state particles (multi-particle state). The final state often consists of many hadrons since the photon-photon collision has coupling either to a quark or a neutral vector meson, which contains a $q\bar{q}$ pair

and positron exist within the acceptance of the detector. The threshold $0.05E_{\text{beam}}$ may vary depending on the experimental condition. The last criterion reduces the probability that the e^+e^- pair is produced in association with some other particles that escape from detection. This requirement would remove effectively events from so-called two-photon processes (see Fig. 2.4) where an e^+e^- pair is produced in the final state and both electron and positron lose significant energies.

You may like to add further criteria to constrain the process to increase the fraction of the process in your mind among the event sample, so that the interpretation of thus defined cross section becomes more intuitive. This kind of event selection is called *exclusive* event selection. The above-given example would be called measurements of exclusive e^+e^- production.

Another way to investigate the underlying physics of e^+e^- collisions through e^+e^- final state is to define the selection criteria as simple enough. An extreme would be to just select an e^+e^- pair, both of which are energetic enough, like

- a pair of electron-like particles in opposite charge, both within the angular region $\theta_{\text{min}} < \theta < \pi - \theta_{\text{min}}$;
- each electron fulfils $E_{\text{elec}} > 0.3E_{\text{beam}}$, where E_{elec} is the energy of the electron.

This allows you to select events which include other particles in the final state; for example, the selected events contain, with a high probability, the process from Fig. 2.4. This kind of event selection is called *inclusive* event selection. The benefit of the inclusive event selection is that it is indeed "well defined" in terms of the theoretical prediction. For measurements with exclusive event selections, the modelling of the soft emission of many particles in the theoretical prediction (Monte Carlo simulation) is often not easy since it involves higher orders of the perturbative calculation. On the other hand, the total production rate, which is one of the measurements with inclusive event selections, is often calculated well since certain techniques exist to sum up all the contributions of final states.

The inclusive measurements are often performed when one would not know the number of particles produced in the reaction, either it is not measured or it is not easy to measure. Examples are jet production in hadron-hadron collisions, e.g. $pp \rightarrow$ $n \times$ jet $+ X$, where $n \geq 1$ and X represents a part of the final state with any number of particles, or in deep-inelastic scattering $eN \rightarrow eX$ where N is a nucleon.

2.4   Event Acceptance and Efficiency

The event selection criteria will reduce the number of events you take from the process of your concern. The remaining number of events becomes smaller if the event selection becomes more exclusive, which is necessary if the amount of contribution from background processes is to be reduced. The solid angle coverage of your detector also gives a hard limit on the detection possibility. All these will cause the reduction in *acceptance*, defined as

$$(\text{acceptance}) = \frac{N_{\text{sel}}}{N},$$

where N_{sel} denotes the number of events passing the selection criteria, imposed on *true* four-momenta of particles, while N is the number of events from the processes in concern. Here, the *true* four-momentum means that it is used in theoretical calculation without detector smearing. This may be available in event generators (see Chap. 7).

The acceptance is often a very small number (like $\ll O(10^{-1})$) if the event selection is exclusive and the denominator is defined as all the events from the process. Instead, the acceptance may become closer to unity if it is defined for differential cross sections at a given point in the phase space both for the denominator and numerator. As an example, for the exclusive event selection criteria given above for the e^+e^- final state, the cross section may be defined as double-differential cross sections $d^2\sigma/dE_1 d\theta_1$, where E_1 and θ_1 are the energy and angle of the highest energy electron. The acceptance for events with either $\theta_1 < \theta_{\text{min}} + \Delta$, $\theta_1 > \pi - (\theta_{\text{min}} + \Delta)$, or $E_1 < 0.8E_{\text{beam}}$ is zero. That means that the differential cross sections corresponding to these kinematic regions are also zero. But for other regions, one would expect the acceptance is closer to unity than the average acceptance over all possible kinematic regions of the processes in concern. In this way, we can avoid a large extrapolation factor from N_{sel} to N.

Since there is no detector with 100% detection efficiency, we necessarily lose events by the inefficiencies of the detectors. We often treat the loss of this effect to the acceptance separately from the geometrical acceptance and call it *efficiency*, defined as

$$(\text{efficiency}) = \frac{N_{\text{det} \cap \text{sel}}}{N_{\text{sel}}},$$

where $N_{\text{det} \cap \text{sel}}$ denotes the number of events passing the selection criteria imposed on quantities obtained from measurements, while N_{sel} is the number of events passing the selection criteria imposed on true four-momenta of particles.

Note that a detailed definition of acceptance and efficiency may be different from what is given above and may depend on physics analysis or literature.

With the acceptance A and efficiency ϵ defined, the number of signal events N_{sig} for collisions with integrated luminosity L_{int} can be derived from

$$N_{sig} = L_{int}\sigma A\epsilon.$$

The cross section can be obtained from this equation.

In most measurements, you cannot ignore the presence of the background events. The N_{sig} should be replaced with $N_{sig} = N_{obs} - N_{bkgd}$, where N_{obs} is the observed number of events and N_{bkgd} is the number of background events. The estimation of background events is time-consuming in data analysis, which is discussed later.

Note that it is often difficult to know L_{int} precisely enough in collider experiments, especially for hadron colliders. Also the deviation in detector calibration from its *truth* value causes shift in the number of observed events, often rather uniformly across other kinematical variables (e.g. angles). The normalisation of a cross section, or differential cross sections, is important for extracting the strength of the interaction such as extraction of coupling constant and higher order effect on perturbation calculation. The normalisation is, however, not necessary when extracting the physics quantities from the shape of the distribution, such as the mass spectra and the spin of the exchanged particles. A normalised distribution is used to extract physics quantities for such purposes.

2.5 Nature of Hadron-Hadron Collisions and Kinematic Variables

Hadron-hadron collisions are realised by two high-energy stable hadron beams brought into collisions. Only protons or anti-protons have been used in modern high-energy accelerators in practice. The proton is not an elementary particle; instead, it consists of partons, i.e. quarks and gluons. The processes that undergo in hadron-hadron collisions are categorised into two: *soft* and *hard* interactions.

In soft collisions (Fig. 2.5a), the constituent of a proton is not resolved during the interaction of, for example, two protons. This occurs when the exchanged particle during the interaction does not carry high momentum. Such an interaction cannot be described perturbatively by quantum chromodynamics (QCD) since the strong

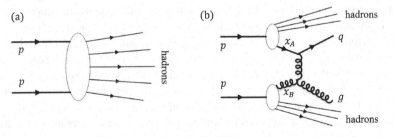

Fig. 2.5 a A schematic view of a soft pp collision with multi-hadronic final state. **b** An example of a hard pp collision with two high-p_T partons in the final state

Fig. 2.6 The strong coupling constant α_S as a function of the renormalisation scale $\mu = Q$. Reprinted under the Creative Commons Attribution 4.0 International license from [2] © 2018 CERN, for the ATLAS Collaboration. The data are taken from measurements from hadron and *ep* colliders. The curve indicates the solution of the renormalisation-group equation using α_S obtained from the results indicated by red (solid circle) points

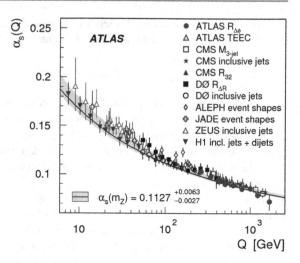

coupling constant α_S is in fact too strong for low-energy interactions. Figure 2.6 shows the behaviour of $\alpha_S(\mu^2)$, where μ represents the energy scale of the interaction, e.g. the four-momentum of the exchanged particle. Since α_S becomes much larger than $O(10^{-1})$ when the momentum transfer is similar to or smaller than $\Lambda_{QCD} \simeq 200\,\text{MeV}$, a perturbative expansion based on the number of partons is no longer possible there. In such a situation, the partons are bound strongly and the nucleons move collectively. Individual partons are no longer visible. We often call such an object "quark matter", like a fluid consisting of quarks and gluons, which binds the quarks together.

The soft interaction would look, therefore, like two pancake-like composite objects moving and crossing across. The collision of such objects may be simplified as follows: the two objects interact with each other if the compound has an overlap with the other composite material, and do not interact if the impact parameter is larger than the diameter of the objects. It is expected that the cross section for such an interaction is constant as a function of the centre-of-mass (CM) energy of the collision, assuming that the size of the proton is approximately independent of the interaction. Measurements at various CM energies show that the cross section rises with energy, but very slowly.

On the other hand, the parton (here denoting the parton A) in a proton *can* resolve the parton (the parton B) of the other proton when the momentum transfer of the exchanged particle is much larger than Λ_{QCD} (see Fig. 2.5b). If the impact parameter of the parton A with respect to the parton B is so small, the field produced by the parton B is strong so that the interaction may occur at a very high energy regime. In other words, the particle exchanged in such an interaction may carry very high momentum with a short wavelength. This allows the partons inside hadrons to be resolved. The α_S gets much smaller such that the interaction can be described by the perturbative calculation. In such *hard interactions*, the lowest order in perturbation, i.e. 2-to-2 interaction becomes dominant; two high-momentum (i.e. *hard*) partons are produced in the final state.

In order to give the kinematic feature of hadronic collisions, let us define the coordinate commonly used for hadron-hadron collisions. We choose the system where the z-axis is along the beam line, with the positive direction of the z-axis being the running direction of the beam A (like the proton beam for $p\bar{p}$ colliders or one of the proton beam rotating counter-clockwise for pp colliders). The x-axis is then chosen towards the centre of the accelerator ring. The y-axis is defined so as to form a right-hand coordinate system. The transverse plane is the xy plane perpendicular to the beam direction. The transverse momentum is defined as $p_{\mathrm{T}} = \sqrt{p_x^2 + p_y^2} = p \sin\theta$ where θ is the polar angle of the coordinate system given above. We often use (p_{T}, ϕ) instead of p_x and p_y, where ϕ is the azimuthal angle.

A soft-interaction event is characterised by an absence of particles with high p_{T}, while at least a few high-p_{T} particles are produced in the hard collisions. The reason to use p_{T} comes from the nature of the hadron-hadron collisions. The hard collision occurs between two partons, whose energies are not equal even if the hadron beam energies are the same. The momentum of the parton A (B) can be expressed as $p_{\mathrm{A(B)}} = x_{\mathrm{A(B)}} p_{\mathrm{beam}}$ using a momentum fraction $x_{\mathrm{A(B)}}$, which is defined as the ratio of the parton momentum involved in the hard collision to the momentum of a hadron to which the parton belongs. In general $x_{\mathrm{A}} \neq x_{\mathrm{B}}$, meaning that the centre-of-mass frame of the two partons involved in the hard collision is boosted against the centre-of-mass frame of the two beams, which corresponds to the laboratory frame for symmetric colliders. Therefore, the only component of the momentum preserved in the hard collision is the one in the transverse plane. The longitudinal component of the centre-of-mass system of the partons A and B cannot be determined unless the values of x_{A} and x_{B} are obtained by other experimental quantities.

Now we may like to determine the third coordinate component p_z of the momentum of the parton-parton collision system. However, it is not always possible to measure the third coordinate precisely in hadron-hadron collisions, since a large fraction of the longitudinal momentum is lost through particles entering in the beam pipe, and some of the lost particles may have emerged from the parton-parton collision. The variable $E - p_z$ of an event, instead, can well be measured even if we lose particles lost in the beam pipe in $+z$, since $E - p_z$ contribution from such particles with very small scattering angle, escaping the beam pipe, is almost zero. Similarly, $E + p_z$ is also well measured even if we lose particles collinear to the $-z$ direction. Constructing variables using $E - p_z$ and $E + p_z$ of a system (event or a part of the event), or a particle, would therefore be determined with certain accuracy. The most common and convenient choice is to use the *rapidity y*, defined as

$$y = \ln \sqrt{\frac{E + p_z}{E - p_z}} = \ln \frac{E + p_z}{m_{\mathrm{T}}}, \quad m_{\mathrm{T}} \equiv \sqrt{m^2 + p_{\mathrm{T}}^2}.$$

A good property of the rapidity is that the difference in rapidity between two four-momenta is preserved under a Lorentz boost. This also means that the Lorentz boost can be calculated by adding the rapidity of the boost vector.

In a limit where a particle is massless, the rapidity is equal to pseudorapidity η, defined as

$$\eta = -\ln\tan(\theta/2).$$

η is a good approximation of y in modern high-energy collider experiments, where the particles from hard collisions are produced at $> O(10)\,\text{GeV}$, which is much higher than typical mass of long-lived final state particles like electron, muon, pion, or kaon. At $\theta = \pi/2$, $d\eta/d\theta = 1$, i.e. $\Delta\eta$ corresponds exactly to $\Delta\theta$.

Since the Lorentz transformation of the rapidity is additive, it expresses well the Lorentz-invariant phase space of final state particles. The phase space for a particle is

$$d^4 p\,\delta(p^2 - m^2) = d^3\text{p}/E = \pi dy dp_T^2\,,$$

where m is the mass of the particle and p is the three-momentum of the particle while p denotes the four-momentum. This means that the particle is uniformly produced in y if the particle is equally distributed in the phase space. A differential cross section is, therefore, expressed often as $d\sigma/dy$ instead of $d\sigma/d\theta$. The latter is more commonly used in non-relativistic collisions, fixed target experiments, or e^+e^- collisions.

2.6   Structure of Hadrons and Parton Density Function

The hard process as introduced in the previous section regards a hadron-hadron collision as a scattering of one parton from the parent hadron A of its momentum fraction $x_A = p_A/p_{\text{beam}_A}$, with another parton from the other parent hadron B, $x_B = p_B/p_{\text{beam}_B}$. For high-energy collisions where the mass of the partons can be ignored, the centre-of-mass energy of the two partons A and B, $\sqrt{\hat{s}}$, is given as $\sqrt{x_A x_B s}$.

In order to estimate the cross section of hard collisions, one needs to know the "luminosity" of such partons, i.e. number of partons in the incoming beam particles, in order to convert the luminosity of hadron-hadron collisions to the luminosity of parton-parton collisions. The number density would depend on x of the parton since a hadron is a composite particle consisting of partons of various momenta. Now a slight complication is that the number of partons with a given x also depends on the wavelength of the probe. This is explained qualitatively as follows.

One needs an electron microscope to see the structure of viruses since the virus is smaller than the wavelength of the visible light. The electron beam energy of the electron microscope ($\gg$ keV) is much larger than that of the visible light so that one can resolve the fine structure of the virus. Similarly, if we like to see the structure of hadrons, it is necessary to use a probing beam, whose wavelength is much shorter than the size of the hadron itself $\simeq 1$ fm $\sim 200\,\text{MeV}$. Practically, what probes the structure of hadrons is not the beam itself but rather a particle coupling to the partons, e.g. photons for the electron beam, or gluons, or quarks for hadron beams.

Now, what is known is that more partons (the structure of hadrons) are seen as the wavelength of probe particles gets shorter, as schematically drawn in Fig. 2.7. In this figure, an electron as a projectile collides with a quark inside a target proton.

Fig. 2.7 A schematic drawing of the $ep \rightarrow eX$ reaction where a γ^* is exchanged between the proton and the electron. The large blob in the left figure indicates the area that can be probed by the virtual photon with a long wavelength. As the wave becomes shorter like in the right figure, more parton radiation may be visible

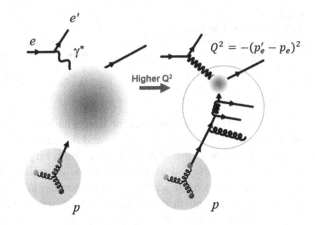

Such an interaction is called deep-inelastic scattering (DIS). The scatter exchanges a virtual photon γ^*, which mediates the force and probes the quarks inside the proton with a short wavelength if $Q^2 \equiv -(p'_e - p_e)^2$ is large, since the wavelength is $\propto \sqrt{1/Q^2}$. When the partons start to feel the external force from the electron beam, an energetic parton may radiate another parton before coupling to the virtual photon, since the coupling constant involved in the parton radiation, α_S, is larger than the electromagnetic coupling. This behaviour of the parton radiation is very well described by a theoretical framework based on perturbative QCD, for example, by the DGLAP equation (for review, see, for example, Ref. [3] or more in detail in Ref. [4]). The equation, with experimental data from lepton-hadron scattering experiments, tells that partons should increase with Q^2, except for very high-x partons, which give momentum to low-x partons through the parton radiation.

As a consequence, the number of partons, or the parton density function (PDF) f_i, where i denotes the type of the parton (gluon or quark flavour), depends not only on x but also on the wavelength of the probe: $f_i = f_i(x, Q^2)$. For low-x(< 0.1) regime, the number of partons increases logarithmically with Q^2. It also shows rapid increase as x of the parton gets lower ($x \ll 0.1$), $f_i \propto x^{-\lambda}$ where λ is typically 0.2–0.4. An example of the parton density functions may be found in the review section for the structure function in the Particle Data Group review [5] and references therein.

For hadron-hadron collisions, a projectile is a hadron and the exchanged force for the scattering is propagated also by a parton. The Q^2 of such collisions needs p_z of the scattered parton, which is not well reconstructed. Instead, p_T^2 of the hard-scattered parton is used as the probing scale when estimating the parton density. The cross section of such hard collisions can be expressed as a product of the parton densities of the partons A and B and the scattering cross section AB $\rightarrow$ CD where the parton indices C and D are the two outgoing partons, as shown in Fig. 2.8:

$$\sigma \propto \sum_{q,g} f_A(x_A, \mu_F^2) f_B(x_B, \mu_F^2) \sigma_{AB \rightarrow CD}(\hat{s}, \mu^2).$$

The probing scale is called a factorisation scale μ_F, which is p_T in this case.

Fig. 2.8 A schematic
diagram showing how hard
scattering cross section is
factorised into

$$f_B(x_B, \mu^2)$$

$$f_A(x_A, \mu^2) \quad D \quad \sigma_{AB \to CD}$$

This formula assumes that a parton from a hadron collides with a parton from the other hadron. This picture may not be valid if the number of partons inside a hadron is very large (in particular if the partons are from very low-x) or the scattering cross section is very large (e.g. due to large α_S in low-p_T regime). For such cases, more than one parton pair may cause scatterings. If the number of scatterings becomes so many, the interaction may not be described anymore by perturbative QCD. They may have to be described, at least partially, by a theoretical framework for soft interactions, which treats the entire hadron as one body for the interaction.

References

1. Aaboud, M., et al.: ATLAS Collaboration. JHEP **1710**, 182 (2017). https://doi.org/10.1007/JHEP10(2017)182. arXiv:1707.02424 [hep-ex], https://atlas.web.cern.ch/Atlas/GROUPS/PHYSICS/PAPERS/EXOT-2016-05/
2. Aaboud, M., et al.: ATLAS Collaboration. Phys. Rev. D **98**(9), 092004 (2018). https://doi.org/10.1103/PhysRevD.98.092004. arXiv:1805.04691 [hep-ex]
3. Salam, G.P.: CERN Yellow Rep. School Proc. **5**, 1–56 (2020). https://doi.org/10.23730/CYRSP-2020-005.1
4. Ellis, R.K., Stirling, W.J., Webber, B.R.: Camb. Monogr. Part. Phys. Nucl. Phys. Cosmol. **8**, 1–435 (1996)
5. Zyla, P.A., et al.: Particle Data Group. PTEP **2020**(8), 083C01 (2020). https://doi.org/10.1093/ptep/ptaa104

Apparatus

3

In this chapter, we present an overview of the apparatus used in high energy physics. First, we describe how the particle collisions with high energy are obtained, and then how the reaction of the particle collisions is recorded as the experimental data.

3.1 Particle Collisions at High Energies

The high energy particle physics (in short, the high energy physics) is the research to reveal the ultimate constituents of the universe and the rules obeyed by elementary particles by experimentally observing particle reactions. Since the size that can be probed is determined by the de Broglie wavelength, $\lambda = h/p$, the availability of the high energy particle is essential to investigate more microscopic worlds. At the same time, momentum transfer produced at particle-particle collisions can be used to generate another particle, which is different from the ones before the collision. This implies that the higher the momentum transfer, or the higher the collision energy, the heavier the particles that are generated. For these reasons, we have been using particles with high energy in the past and we will in the future. The increase of the collision energy is the history of the high energy physics.

In the ancient days of the high energy physics, only cathode rays or particles emitted from radioactive materials are available as a source of the particles for research. The naming convention of $\alpha, \beta, \gamma, \ldots$, reflects such history. As time went by, physicists discovered cosmic rays and started to use them as a source of particles. This is still widely used in modern high energy physics. Neutrino experiments underground are typical examples; there are so many facilities to detect and study cosmic, solar, and atmospheric neutrinos. In the meanwhile, physicists succeeded in building accelerators that allowed them to study artificially produced subatomic particles. Many new particles were discovered and investigated by accelerator experiments.

© The Author(s) 2022
K. Hanagaki et al., *Experimental Techniques in Modern High-Energy Physics*,
Lecture Notes in Physics 1001, https://doi.org/10.1007/978-4-431-56931-2_3

The main subject of this book is, however, the accelerator experiments, or more specifically the collider experiments. Below, we concentrate on the topics of the collider experiments: accelerators and detectors.

3.2   Accelerator

There have been various types of accelerators in the world. Here, we describe a large hadron collider (LHC) [1] at CERN as an example of a large accelerator complex. LHC consists of several different accelerators shown in Fig. 3.1.

The LINAC is the first accelerator that accelerates protons or actually H^- to 150 MeV. The electrons of H^- are stripped off just before the injection to the Booster, and H^- become the proton. The Booster accelerates protons to 1.4 GeV and sends them to proton synchrotron (PS) where the protons increase their energy until 25 GeV. The protons are further accelerated by super proton synchrotron (SPS) to 450 GeV, and then finally injected into LHC. The proton energy can be increased to 7 TeV in the design of LHC, but the largest achieved energy so far is 6.5 TeV as of writing this book (in 2021). As we have just seen LHC as an example, it is very common that the large collider complex consists of several accelerators.

The beam energy is one of the most important parameters in the collider experiments, which is related to the potential to generate heavy particles, the interaction cross section, and so on. It is a long-standing tradition that the physicists have been looking for something new in the particle reaction initiated by the highest energy collisions. In fact, the history of discoveries in the high energy physics is the his-

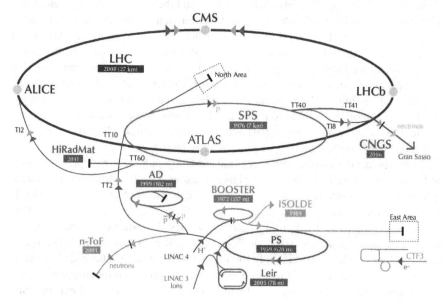

Fig. 3.1 CERN accelerator complex [1]. Reprinted under the Terms of Use from [2][2] © 2013-2022 CERN. All rights reserved

tory of the accelerators where the beam energy has been increased. Higher energy machines have enabled us to "see" the subatomic world with higher resolution, and generated heavier objects, such as the Higgs boson.

On top of that, the luminosity of the particle collisions is another key parameter of the experiment. The higher the luminosity, we can accumulate more data per unit time. This allows us to improve the precision in the view of statistical uncertainty, or to search for rarer events such as particle decays with small branching fractions.

Particles inside the synchrotron such as LHC are accelerated by radio frequency waves (RF) that are generated by a so-called RF cavity. Therefore, only the particles that are located in an appropriate phase of the RF can be accelerated. If not, they are decelerated and cannot be in the orbit of the accelerator. We call the cluster of particles spaced by the RF a "bunch". LHC is operated with the 40 MHz bunch frequency, and hence the bunch crossing occurs every 25 ns if all the bunches are filled with protons.

While explaining above, we have paid careful attention, i.e. we have properly used two terms: the bunch crossing and the collision. The bunch crossing means that two clusters of particles cross at a small space region; some particles are interacted, which is the particle collision. In the LHC, that is, proton-proton collisions, if the number of protons in each bunch is small, or the proton beams are not squeezed enough, particle collisions would not occur so frequently, although the frequency of the bunch crossing is 40 MHz. Such a situation is said to be a "low luminosity". There is a different story: the cross section in electron-positron collisions is much lower than that in proton-proton collisions, and hence the particle collisions may not occur at every bunch crossing even with the high luminosity electron-positron collider. This is true for the KEKB experiment, for example. In the LHC, however, the bunch crossing is almost identical to the proton-proton collisions. Let's discuss a concrete example. Assuming the proton-proton collision cross section to be 80 mb and the instantaneous luminosity of 2×10^{34} cm^{-2}s^{-1}, the number of collisions or interactions per unit time is 16×10^8. Let's also assume that this luminosity is achieved with the 25 ns bunch spacing, which is close to the case in the actual LHC running in 2018. Based on these assumptions, the number of average interactions per bunch crossing is $16 \times 10^8 \times 25 \times 10^{-9} = 40$. In reality, the number of protons are not uniformly spread across the bunches. Also, there is a statistical fluctuation from the average value. But one can imagine that it is very rare to have zero interactions for a bunch crossing. Therefore, we will use the words "bunch crossing" and "particle collision" or "particle interaction" with the same meaning later on if there are no confusions.

So far, we have just discussed the colliders. In addition, there are other types of accelerator experiments, the fixed target experiments. As shown in Fig. 3.1, for example, the Booster, PS, and SPS are used for various fixed target experiments. Particles accelerated by the accelerators are extracted and injected to a fixed target, instead of being collided with each other. With these types of experiments, it is much more difficult to increase the centre-of-mass energy than the colliders, but it is much easier to have high rate interactions at the target due to the large size of the target.

For this reason, the fixed target experiments are suitable for getting high statistics, and widely used for rare decay searches.

There is a concept of "spill" at the fixed target experiment, which doesn't exist in the collider. In the case of the collider experiments, the particle-particle collisions last until the luminosity becomes low because of the beam lifetime. On the other hand, all the particles inside the accelerator are extracted in a certain amount of time, the order of seconds or minutes, at the fixed target experiments. Once all the particles are extracted, new particles are injected through the injector chain to the main accelerator. This cycle is repeated in the fixed target experiments. Therefore, the beam is only available for an experiment when the particles are extracted and hit the target. This period is called "spill".

3.3   Detector

The particle collisions induced by the collider, fixed target, or cosmic-ray experiments need to be captured by some means. Many particles are produced in these collisions. Some people look for new particles, new decay chains, new patterns in the event kinematics, and so on, which are not discovered yet. Others try to measure the rates of specific reactions such as cross sections or branching ratios of particles. In any case, we want to detect all particles produced by particle reactions and to measure their trajectories, and energies (and flight times if necessary) as precisely as possible. In this regard, geometrical acceptance, detection efficiency, and resolution on measurements are the important figures of merit in considering detectors.

Below, let's take a close look at $t\bar{t}$ pair-production events, where a pair of the top and anti-top quark is produced, as an example to see what we have to detect and measure in high energy experiments. As the top quark immediately decays to b-quark and W boson with the probability close to 100%, a $t\bar{t}$ pair becomes two pairs of b and W without leaving any trace of top quarks in detectors. The W boson decays to $e\nu_e$, $\mu\nu_\mu$, or $\tau\nu_\tau$ with the probability of about 11% each, and a quark-anti-quark pair with the probability of about 66%. The former is called a leptonic decay and the latter a hadronic decay. This results in three types of final states.

- Both W decay leptonically, called a dilepton or two-lepton channel.
- One W decays leptonically, and the other hadronically, called a lepton+jet or one-lepton channel.
- Both W decay hadronically, called an all-hadronic or no (or zero)-lepton channel.

We use the lepton+jet channel as a further example in order to describe the particle detection, because the variety of particles in the lepton+jet channel is more than that in the other two channels.

Here, let's assume one W decays to $e\nu_e$ or $\mu\nu_\mu$, and the other hadronically. Then the $t\bar{t}$ final state consists of two b-quarks, one electron (muon), one ν_e (ν_μ), and two more light (u, d, c, or s) quarks. The experimentalists want to detect all these particles, and hence try to make the detector more hermetic, i.e. the larger solid angle coverage with respect to the particle interaction point is more preferable. The next question is how to detect and identify all these particles. In the following, we provide an overview of the basics of how each type of particle interacts with materials or detectors, and how they are captured. The detail of the particle identification will be discussed in Chap. 6.

3.3.1 Particle Interaction with Material

3.3.1.1 Electron and Photon

Electrons with the energy of our interest create electromagnetic showers immediately after hitting dense materials such as calorimeters. Hence, the energy and the position can be measured at calorimeters by sensing the energy deposit of electromagnetic showers. At the same time, most of the detectors used in high energy physics have a device to measure trajectories of charged particles, which allows us to measure the momentum in conjunction with the magnetic field provided by a magnet. This is called a magnetic spectrometer. In addition, the precise tracking of charged particles provides the information to find a particle collision point in the collider experiments, and helps particle identification which will be described later in detail.

In addition, photon is a very similar object to electron in terms of detection in high energy regime because photon hitting a material also creates electromagnetic shower. So electromagnetic calorimeter usually measures the energy of both electrons and photons. But there is an important difference, i.e. photon is a neutral particle and hence no track is detected with the charged particle tracking system. This difference is actually used to distinguish photon from electron.

3.3.1.2 Muon

Because muons with their momenta under our interests do not make electromagnetic showers in materials, their energy deposits are almost only by the ionisation of detector materials. This feature allows us to discriminate muons from other charged particles by placing enough materials, which are usually a part of the detectors such as calorimeters and/or solenoid magnets. In most collider experiments, there are two charged particle trackers, one located near the particle collision point, and the other after dense materials such as calorimeters. Particles detected after the dense materials can be identified as muons with a high probability. By connecting trajectories measured by the two trackers, one can assure the muon really comes from the particle collision point.

3.3.1.3 Quark (or Gluon)

Any quarks produced by particle collisions or decays from the other particles are immediately hadronised, except for top quarks, because of the feature of QCD (see Sects. 2.5 and 6.4.1). In the $t\bar{t}$ events, there are two b-quarks produced by the decay of top quarks, and two light quarks decayed from W. All four quarks are metamorphosed to hadrons. The number of hadrons that emerged from a single quark mostly depends on the energy of the original quark. As higher the energy, more hadrons are emerged. Because the top quarks or W bosons are much heavier than light quarks, many ($O(10)$) hadrons are formed for each of the four quarks, which are aligned to the direction of the momentum vector of the original quarks. This cluster of such particles is called a jet, which will be explained in Sect. 6.4.

It would be ideal to measure the momenta of all the particles inside a jet. However, the magnetic spectrometers cannot detect neutral particles such as photons and neutrons. Because a π^0 meson immediately decays to two photons with a branching ratio close to 100%, two photons need to be detected. This fact leads us to a tradition that the energy and direction of jets are measured by calorimeters. More specifically, hadrons are measured by a combination of electromagnetic and hadronic calorimeters behind, in contrast to electromagnetic showers such from electrons and photons that are detected by electromagnetic calorimeters.

3.3.1.4 Neutrino

The cross section of neutrinos to interact with materials is too low to detect. Except for dedicated facilities for neutrino experiments, the nominal collision experiments cannot detect neutrinos, causing "missing energy". In the electron-positron symmetric-energy colliders, for example, the momentum and energy of the initial states is well defined, i.e. the sum of momenta is zero. The momentum conservation allows us to deduce the momentum vector of neutrinos from the missing momentum, assuming the detector is hermetic enough.

One needs to modify the above idea slightly for hadron colliders. A proton consists of many quarks and gluons, i.e. partons, in the picture of the high energy physics. What actually collide with each other in proton-proton colliders, for example, are partons in protons, not protons themselves. This means that even at symmetric-energy hadron colliders, the actual energy used for a collision is asymmetric, because the net energy of colliding partons varies event-by-event, and there is no principle or law that forces two colliding partons to have the same energy. Humankind does not predict which partons actually collide with each other and how large energy they have event-by-event basis, even though we can know the momentum of the protons. Therefore, the momentum of the beam direction cannot be used at the hadron colliders. The momentum conservation law can be used only for the plane perpendicular to the colliding beams. Here, we ignore the Fermi motion of the partons inside protons because its energy is negligibly small compared to the colliding beam energy. Thus at hadron colliders, neutrino momenta can be measured only on the plane perpendicular to the beam, called "missing p_T" or "missing E_T", which could be a vector (x, y components) or a scalar (the magnitude of a vector) depending on the context.

3.3.2 ATLAS Detector

As we have just seen what kinds of particles and what properties need to be detected, we next discuss how they are detected. The layout or configuration of a multi-purpose detector for high energy physics is common for many experiments, because if you want to detect all kinds of particles, the layout would become unique based on the nature of the interaction of each particle. The most inner part is covered by a charged particle tracker with the material as low as possible so that all particles can penetrate the tracker and reach calorimeters for energy measurements. Because radiation length is much shorter than interaction length, i.e. an electromagnetic shower evolves much faster than a hadronic shower, an electromagnetic calorimeter is placed in front of a hadronic calorimeter. A muon is identified by the fact that it is rare to make either electromagnetic or hadronic shower in our energy region, and hence it penetrates through massive materials such as the calorimeters. Therefore, a muon detector is located on the outermost part of a whole detector system. To summarise, the order of detector elements tends to be a charged particle tracker, an electromagnetic calorimeter, a hadron calorimeter for energy frontier experiment, and a muon detector from inside to outside.

Since the concept is common for most of the detectors, we use the ATLAS detector [3] in the following as an example to introduce the actual detector. Figure 3.2 shows the ATLAS detector consisting of a barrel and two endcap parts. Each barrel and endcap is actually a collection of various detector components, which will be described later. There is a beam pipe penetrating the middle of the detector to make the proton beams run through it. In addition, there are Solenoid and Toroid magnets to provide a magnetic field, allowing to measure the momentum of charged particles. The Solenoid locates between the charged particle tracker and the electromagnetic calorimeter, and the Toroids outside the hadron calorimeter, covering high-$|\eta|$ regions. The field strength by the Solenoid is 2 Tesla. The integrated field strength by the Toroid varies from 2 to 9 T·m depending on $|\eta|$ and ϕ.

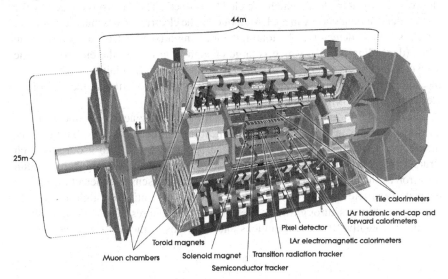

Fig. 3.2 Overview of the ATLAS detector [3]. Reprinted under the Terms of Use from [4] ATLAS Experiment © 2008 CERN. All rights reserved

Surrounding the proton-proton interaction point is the charged particle tracker consisting of the pixel and strip-type silicon detectors; each is referred to as pixel and semiconductor tracker (SCT), respectively. All the charged particles such as electrons, charged pions, muons, and so on interact with the tracker materials, and lose their energy, resulting in the creation of electron and hole pair inside the silicon sensor. These holes and/or electrons are collected by the electric field inside the sensor to the electrode, amplified, and recorded as the signal of the particle hit. The pixel and strip detectors have many layers of sensors, enabling us to "reconstruct" the particle trajectory by connecting the space hit points in many layers.

Outside the silicon detector is another tracking device of charged particles, consisting of many transition radiation tubes, referred to as transition radiation tracker (TRT). The mechanism to detect charged particles is similar to the silicon detectors. Each tube of TRT is filled with gas which acts as the sensor instead of silicon. The charged particles passing through the gas create ion and electron pairs that are read out as a signal through the electrode, either cathode or anode wires. In the case of the silicon detectors, usually they have fine pixel or fine pitch of the strips to achieve good space resolution, typically the order of 10 or 100 μm. On the other hand, the gas-based tracking device such as TRT uses time information on top of the discrete hit information collected by wires. By knowing the drift time of the ions and/or electrons in the gas, one can deduce more precisely the location of the particles interacting with gas by recording the time of signal arrival. Although the typical size of the tube is the order of mm, $O(100\,\mu\text{m})$ position resolution can be achieved.

Most of the charged and neutral particles penetrate the tracking detectors, and hit into the electromagnetic calorimeter composed of the sandwich structure with lead and liquid argon (LAr). Electrons and photons develop the electromagnetic showers mainly at the lead, which is called "absorber". The electrons created by the shower deposit their energy in the LAr, inducing the electric signals that are recorded, which is called "detector". The total radiation length (see Sect. 6.2.1) is more than $24X_0$ (depending on $|\eta|$), which is large enough to terminate the electromagnetic showers, leading to precise measurements of the energy. In addition to the energy measurement, the segment of the calorimeter allows us to identify the location of the electrons or photons hitting into the calorimeter.

The hadron calorimeter is located outside the electromagnetic calorimeter. There are some varieties in the detector types depending on their locations, but the common concept, also used for the ATLAS hadron calorimeter, is to use a sandwich structure made from the absorber and the active region (detector). The barrel region uses iron as the absorber, and plastic scintillators as the sensor to detect the energy deposit of particles created by the hadron showers. A scintillation light is detected by photomultipliers through the wavelength shifting fibres. The total interaction length[1] is roughly 10 λ_0. Only muons and neutrinos in the SM can penetrate the

---

[1] The (nuclear) interaction length λ_0 (λ_I, λ_{int}, etc.) is a useful parameter for the hadron showers, which represents the mean free path for inelastic collisions. The basic idea is similar to X_0. In practice, λ_0 can be expressed by $35A^{1/3}$, where A is the atomic number and its unit is g/cm^2 or cm: λ_0 for iron is 17 cm. Typical hadron calorimeters have about 10 λ_0.

hadron calorimeters except for punch-through hadrons. In case particles could not be stopped by the calorimeters, that is, their showers are leaked behind, such particles could be detected by other detectors (practically muon detectors). Such particles are called "punch-through" ones (punch-through hadrons).

Following the hadron calorimeter, the outermost layer of the ATLAS detector is the muon spectrometer consisting of monitored drift tube (MDT) and cathode strip chamber (CSC) for precise tracking, and resistive plate chamber (RPC) and thin gap chamber (TGC) for providing fast signal for triggering. These all are the gas detectors like TRT, which allow us to measure the particle passage. The position resolution of MDT and CSC is the order of 100 μm. On top of providing fast signals to form a trigger, RPC and TGC determine the event timing, which means that these detectors resolve in which bunch crossing the interaction occurs. Thus, the timing resolution is required to be high for these detectors.

3.3.3 Trigger

The total inelastic cross section of the proton-proton collisions is about 100 mb at $\sqrt{s} = 14$ TeV in the LHC. When the instantaneous luminosity of the LHC accelerator is reached at 2×10^{34} cm^{-2}s^{-1}, the rate of the inelastic proton-proton interaction is expected to be about 2 GHz. Since the frequency of the bunch crossing in the LHC is designed as 40 MHz, we expect 50 proton-proton collisions in every bunch crossing as discussed in Sect. 3.2 that is called pile-up events. As the instantaneous luminosity goes up with the fixed rate of the bunch crossing, the number of the pile-up events increases more. On the other hand, the event rates of physics of interest, such as the production of the Higgs boson, are expected to be the order of 1–10 Hz or much less, depending on physics processes, as shown in Table 3.1. Thus, the inelastic cross section is huge so that even if events are produced from interesting physics processes, they are overlapped with lots of pile-up events.

The total number of channels of the ATLAS detector is about 2×10^8. The detector sends 40 MHz $\times$ $2 \times 10^8 \simeq 10^{16}$ bits $\simeq 10^{15}$ bytes (1 Peta bytes) data every second, in case each of the channels sends a binary digit every collision. Although the data size per event can be reduced by a factor of about 100 using the noise-like data suppression and a bunch of zero-data suppression techniques, it is still inefficient to record all data of the proton collisions into the data storage system. Before accumulating data of an event into the data storage, its event is analysed online and a decision is made whether or not to keep the event for later offline study. This process is called "trigger". The current ATLAS trigger and data acquisition (DAQ) system is based on two levels of online event selection, called level 1 trigger (L1 trigger) and high level trigger (HLT), respectively, as shown in Fig. 3.3 [5].

3.3.3.1 Level 1 Trigger
The L1 trigger makes an initial selection based on a huge amount of electronic modules (printed circuit boards equipped with application-specific integrated circuits

Table 3.1 The rough cross section and the event rate for typical processes. The centre-of-mass energy and the instantaneous luminosity in the LHC are assumed to be 13 TeV and 2×10^{34} cm^{-2}s^{-1}, respectively

Process	Cross section	Event rate
W boson production	190 nb	3.8 kHz
Z boson production	60 nb	1.2 kHz
Top quark pair production	850 pb	17 Hz
Higgs boson production	60 pb	1.2 Hz

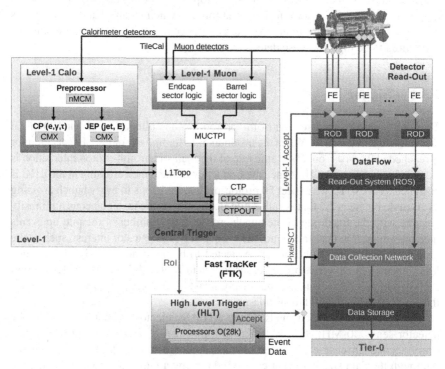

Fig. 3.3 ATLAS trigger and DAQ scheme. Reprinted under the Creative Commons Attribution 4.0 International License from [5] © CERN for the benefit of the ATLAS collaboration 2017

(ASICs) and field-programmable gate arrays (FPGAs), forming multi-chip modules) and their interconnections using information with reduced granularity as inputs from a subset of detectors. There are two main L1 trigger systems in ATLAS. The L1 calorimeter trigger (Level-1 Calo in Fig. 3.3) is based on reduced-granularity information from electromagnetic and hadronic calorimeters, and searches for the high p_T electrons and photons, jets, and taus decaying into hadrons, as well as large missing and total transverse energy. The L1 muon trigger (Level-1 Muon) is based on information from so-called trigger chambers; resistive plate chambers (RPC) in the barrel and thin gap chambers (TGC) in the endcaps, and selects high p_T muons.

The number of objects such as muons, electrons and photons, jets, and taus above the set of threshold of p_T or E_T (for example, the threshold of the muon momentum is set as 6, 10, 15, 20 GeV, and so on) in the fiducial region are counted and sent to the L1 central trigger processor (Central Trigger). The L1 trigger provides "region-of-interest (RoI)" information including position (η and ϕ) and p_T range of candidate objects for the input of HLT. In the case of the trigger based on the missing and total transverse energy, the information on whether an event passes through the criterion of the threshold is sent to the L1 central trigger processor. The central trigger processor makes an L1 trigger decision based on the combination of objects required in coincidence or veto and provides the signal of the "L1 accept" (Level-1 Accept). The L1 trigger makes a trigger decision within about 2.5 µs and reduces the event rate from 40 MHz to 100 kHz. During the process of the trigger decision, information for all detector channels has to be retained in "pipeline" memories, which are placed on usually front-end electronics systems of the detectors (FE in Fig. 3.3). The depth of the pipeline memories depends on the size of data per event, the frequency of the trigger latency.[2]

3.3.3.2 High Level Trigger

Only events selected by the L1 trigger are read out from the front-end electronics systems to the readout systems (ROS). Further trigger selections are done by the HLT. The HLT makes a more precise selection based on a huge amount of processors. Using the RoI information, the HLT selectively accesses data from readout systems. Typically, only data from a small fraction of the detector, corresponding to RoI information provided by the L1 trigger, are needed by the HLT. Hence, usually only a few per cent of the full event data are required for the event processing. The HLT makes use of information from muons, electrons, photons, jets, taus decaying into hadrons, missing and total transverse energy, and the charged particle tracks provided by the inner tracking system. More specifically, combination of p_T or E_T of the objects above, and topologies of events such as invariant mass and angles between the objects are used for a decision of the HLT. Only events accepted by the HLT are recorded in the data storage. The HLT reduces the event rate from 100 kHz to a few kHz.

3.3.3.3 Trigger Requirements for Selecting Physics Events

The trigger should reduce the data while keeping candidate events for further physics analyses. The target physics can be the SM process including the production of Higgs, W and Z bosons, and searches for signatures beyond the SM such as supersymmetry or other theoretical models. The trigger needs to cover all signatures for these target physics processes using electrons, photons, muons, jets, taus, b-jets, and missing transverse energy. A few thousands of different trigger conditions are prepared, and the list of these triggers is called a "trigger menu". The trigger menu is frequently

---

[2] Thanks to the progress of the high-speed optical transmitter, hit information from all detector channels can be transmitted to the electronics modules on a counting room.

updated depending on the accelerator conditions and physics of interest. Practically, before starting your physics analysis, you need to design the trigger condition to store events of your interest adequately while keeping the trigger rate of background events low enough.

For example, the candidates of Higgs production followed by the decay of $H \rightarrow ZZ^* \rightarrow \mu\mu\mu\mu$ can be collected by a combination of the L1 muon trigger with $p_T > 15$ GeV threshold and the HLT with $p_T > 20$ GeV threshold. In this case, the trigger efficiency is high enough for muons reconstructed to be really above the "turn-on curve" i.e. $p_T > 15$ GeV for L1 and $p_T > 20$ GeV for HLT, while the efficiency is low if the muons are well below these thresholds (Fig. 3.4). If at least one muon out of four muons from Higgs decay passes through the fiducial detector volume and has $p_T > 20$ GeV, this kind of event can be kept for later physics analysis. Background events from the inelastic proton-proton interaction with a lot of low p_T particles, mostly hadrons, may be effectively rejected by the muon trigger with the high p_T threshold. However, there are background events that are not removed by the trigger, where a charged hadron is misidentified as a muon, a low p_T muon is mismeasured as a high p_T muon, or a few low p_T tracks are combinatorially reconstructed as one high p_T muon. As discussed at the beginning of this section, since the cross section of the inelastic proton-proton interaction is very high compared to that of the interesting physics processes in most cases, the trigger rate can be dominated by background

Fig. 3.4 The efficiency of the muon trigger. Reprinted under the Creative Commons Attribution 4.0 International License from [5] © CERN for the benefit of the ATLAS collaboration 2017. Top and bottom show the trigger efficiencies for the barrel region and the endcap region, respectively. The efficiency of the barrel region is lower, because in some regions it is hard to place the muon chambers due to the interference of the toroidal magnet

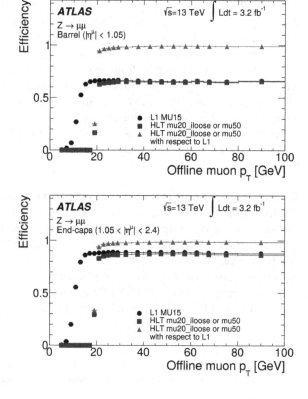

events even though the misidentification and the mismeasurement of muons are rare. The trigger rate needs to be monitored as a function of the instantaneous luminosity shown in Fig. 3.5 and controlled by optimising, for example, the threshold for p_T of the objects in concern.

Fig. 3.5 The rate of the muon trigger. Reprinted under the Creative Commons Attribution 4.0 International License from [5] © CERN for the benefit of the ATLAS collaboration 2017

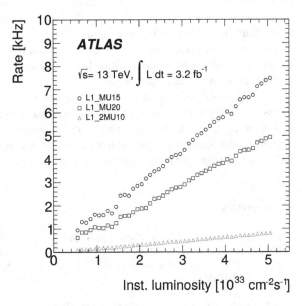

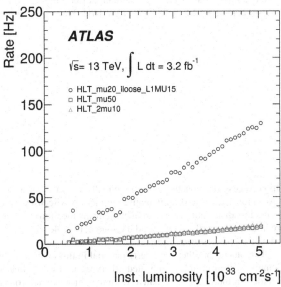

3.3.4 Optimisation of Detector Performance

If the frequency of the event that needs to be recorded is low, the detector can be optimised for its resolution and/or efficiency. In order to achieve high position resolutions, for example, one might decrease the size of each pixel in the pixel detector for the tracking. However, this increases the number of channels to be read out, and might limit the DAQ speed, which should be improved if necessary. Therefore, the optimisation and compromise are necessary when designing a detector, and their balance depends on many constraints, for example, physics requirements, detector technologies, and some from budgets.

The readers should be aware of the fact that not only detectors but also experiments themselves are strongly constrained by such boundary conditions in reality. It would be instructive to think about or imagine the constraints which are imposed on the detector under study, and why such a particular design was chosen. Such training will help to design and build your own detectors and experiments.

References

1. Evans, L., Bryant, P.: JINST **3**, S08001 (2008). https://doi.org/10.1088/1748-0221/3/08/S08001
2. https://cds.cern.ch/record/1621583
3. Aad, G., et al.: ATLAS collaboration. JINST **3**, S08003 (2008). https://doi.org/10.1088/1748-0221/3/08/S08003
4. https://cds.cern.ch/record/1095924
5. Aaboud, M., et al.: ATLAS collaboration. Eur. Phys. J. C **77**(5), 317 (2017). https://doi.org/10.1140/epjc/s10052-017-4852-3. arXiv:1611.09661 [hep-ex]

Statistics

4

Statistics is a tool to show how accurate a measurement is. Particle physicists make use of statistics to know, for example, with how much probability the events under study are signal-like in a new particle search, how much improvement a new measurement gives compared with the past measurements, how powerful a new analysis method is, and how much a certain theoretical prediction is restricted. Statistics is also useful to discriminate the signal events from the background events and to estimate the number of the signal events and background events properly. Therefore, the basics of statistics are essential for experimentalists.

As an example of how the statistics is used in particle physics, let us explain how the cross section of a certain physics process (σ_{phys}) can be measured. As described in Chap. 2, σ_{phys} can be extracted from five experimental observables: the number of observed events (N_{obs}), the number of estimated background events (N_{bkgd}), the acceptance of the event selection (A), the detection efficiency (ε), and the integrated luminosity (L_{int}) using

$$\sigma_{\text{phys}} = \frac{(N_{\text{obs}} - N_{\text{bkgd}})}{L_{\text{int}} A \varepsilon}. \tag{4.1}$$

N_{obs} is measured by counting the number of events after the signal event selections. N_{bkgd} is estimated from data and/or the Monte Carlo simulation samples (MC samples). In case we use data, it is often estimated from the fit to a distribution of a physics observable such as an invariant mass of particles. The geometrical acceptance and the detection efficiency of the signal events are often determined using a large amount of MC samples. Signal and background separation can be improved by the selections based on the likelihood method or the multivariate statistical analyses. In the above example, there are five observables and four of them (N_{bkgd}, L_{int}, A, and ε) have both statistical and systematic uncertainties but N_{obs} has only the statistical uncertainty. Systematic uncertainty is an uncertainty that arises from methods performed

© The Author(s) 2022
K. Hanagaki et al., *Experimental Techniques in Modern High-Energy Physics*,
Lecture Notes in Physics 1001, https://doi.org/10.1007/978-4-431-56931-2_4

in the detector calibrations, the data analyses, etc. In the end, the total uncertainty of the cross-section measurement is determined by propagating and combining the statistical and systematic uncertainties of the five observables.

All of what is mentioned here require a good knowledge of statistics. This chapter describes the basics of statistics, which include uncertainties, the probability of the distributions, the propagation of the uncertainties, basic techniques of the fit to a distribution, and the basics of the maximum likelihood method.

4.1   Uncertainty

One can never know the true values of nature, but believe that there are such true values. This idea comes from a kind of frequentist's viewpoint and is adopted in many analyses of collider physics. All we can obtain is the estimator for the true value based on the outcome of the experiment, i.e. measurement. Since perfect measurements can never be performed, the result of measurements is always represented by a centre value and its uncertainty. The centre value is often determined by the most probable value or the expectation value of the measurement. The spread of the probability distribution for the estimator of the central value is often used as the uncertainty, which usually consists of two kinds: the statistical uncertainty and the systematic uncertainty. Therefore, the measurement is usually expressed by

$$\text{(measurement)} = \text{(central value)} \pm \text{(stat. uncertainty)} \pm \text{(syst. uncertainty)}.$$
$$(4.2)$$

In this section, the basic concepts of the uncertainties are explained using the example of the cross-section measurements shown in Eq. (4.1).

4.1.1   Statistical Uncertainty

The statistical uncertainty arises from stochastic fluctuations of random processes. If an event observed is uncorrelated with events observed in the past and the future, the statistical uncertainty follows adequate probability distributions. N_{obs} obeys the Poisson distribution (or the normal distribution if N_{obs} is large enough), which is explained in the next section. The acceptance (A) and efficiency (ε) follow the binomial distribution, which is also explained in the next section. The mean and the uncertainty can be obtained from these probability distributions.

4.1.2   Systematic Uncertainty

Suppose charged leptons are selected in measuring the cross section of a certain physics process. We must know the selection efficiency of the charged leptons to extract the cross section (Eq. (4.1)). The event selection efficiency, which includes all the efficiency of each selection stage, is usually estimated from MC samples.

Ideally, the efficiency obtained from MC samples is the same as the one in real experimental data (in short, real data). But who knows if it is true? In the real life, it is impossible to have the perfect correspondence between MC samples and real data.

The differences must be evaluated, and corrected if needed. Let's keep using the physics process selected by the requirement on charged leptons, as an example, where the selection efficiency is estimated from purely MC samples (defined as ε^{MC}). We must know the selection efficiency of the leptons in real data not in the MC samples. We correct the efficiency of MC simulation if ε^{MC} is not consistent with the one in real data. This correction can be obtained using an under-controlled data and/or well-known physics process (so-called control data), which are collected by the event selection without the lepton requirements. The caveat here is that we cannot use the selection which we want to evaluate when collecting the control data. For example, $Z \rightarrow \ell\ell$, where ℓ represents charged lepton, can be selected by requiring the invariant mass of two charged tracks to be around 90 GeV. Assuming the purity of this control data is high enough for brevity, we can extract the efficiency of selecting the lepton for both MC simulation (ε_{cont}^{MC}) sample and real data ($\varepsilon_{cont}^{data}$).

Once we obtain $\varepsilon_{cont}^{data}$ and ε_{cont}^{MC}, a so-called Scale Factor (SF) is defined as SF $= \varepsilon_{cont}^{data}/\varepsilon_{cont}^{MC}$. Using SF, the efficiency used in Eq. (4.1) is corrected as $\varepsilon = \varepsilon^{MC} \times$ SF. The uncertainty of SF, which comes usually from limited statistics of the control data, is taken into account as the systematic uncertainty of the cross-section measurement. In addition, if ε^{MC} is different from ε_{cont}^{MC} (indication that the lepton selection efficiency depends on physics processes), the difference must be also taken into account as another systematic uncertainty.

Here has been shown one typical example. Usually, we need to consider several different types of systematic uncertainties. The sources of the systematic uncertainty are, for example, poor understanding of jets, electrons, muons, and charged tracks reconstructions, mismodelling of the fitting function to discriminate the background events from the signal events, imperfectness of the theoretical model in the Monte Carlo simulation, and mismeasurement of the luminosity. The study of the systematic uncertainty is essential not only to estimate proper uncertainty of the measurement but also to help understand the details of detector responses and the dependence of what we are measuring on the other physics parameters. The study of systematic uncertainty sometimes also tells us the weakness of the analysis procedure.

4.2 Probability Distribution

Let's assume you roll a dice which is truly a cube. The number n, which is an integer from 1 to 6, will be shown randomly and one can only predict the probability of the number on the dice to be n, which is denoted by $P(n)$. In this case, the probability of n is the flat distribution of $P(n) = 1/6$.

Now further assuming that you throw two cubic-dice, the sum of the numbers on two dice is the random integer from 2 to 12. The probability of $P(n)$ that the sum of the numbers on two dice is n is obtained as a function of n as shown in Table 4.1.

Table 4.1 The probability that the sum of the number on two dice is n

n	2	3	4	5	6	7	8	9	10	11	12
$P(n)$	1/36	1/18	1/12	1/9	5/36	1/6	5/36	1/9	1/12	1/18	1/36

We cannot say what n happens in the next throw but know how large the probability is for each n.

Similarly, in collider physics, it is impossible to predict what kind of event (physics process) will happen in the next collision. A human being can only know the probability of having a certain event in the next collision. The number of observed events (n) in a fixed number of collisions follows a certain probability distribution ($P(n)$).

Not only the number of events but also many other observable quantities such as energy or angle of a particle created by particle collisions, invariant mass reconstructed from particles, and so on follows certain probability distributions. Typical probability distributions which are often used in the particle physics experiments are introduced in the following subsections.

4.2.1 Basics of Probability Distributions

In the example of the dice described previously, the random number n is discrete and $P(n)$ is the probability to have n. For the discrete probability distribution such as the dice, the probability of having from n_j to n_k is given by $\sum_{i=j}^{k} P(n_i)$, if all possible random numbers are distributed as $n_0, n_1, n_2, \ldots, n_j, \ldots, n_k, \ldots n_\ell$. If x is continuous, $P(x)$ is a probability density function of x but is simply called a probability. The probability having x in the interval from x to $\mathrm{d}x$ is given by $P(x)\mathrm{d}x$.

The probability distribution is normalised to 1, so that the probability is defined to be between 0 and 1. If n_i is discrete,

$$\sum_{i=0}^{k} P(n_i) = 1 \tag{4.3}$$

or, if x is continuous,

$$\int P(x)\mathrm{d}x = 1, \tag{4.4}$$

where the integral is over all eligible x.

The expectation value (μ) and the standard deviation (σ) are often used as the measured value and its statistical uncertainty, respectively, in particle physics. So the result of a measurement is usually represented as $\mu \pm \sigma$.

The expectation value is obtained with the arithmetic average. The expectation value is defined to be

$$\mu = \sum_{i=0}^{k} n_i P(n_i) \tag{4.5}$$

if n_i is discrete, or

$$\mu = E[x] = \int x P(x) dx \qquad (4.6)$$

if x is continuous (the integral is over all eligible x).

The standard deviation is the amount indicating how a certain measurement varies statistically from the average. The square of the standard deviation is called the variance and defined as

$$\sigma^2 = \sum_{i=0}^{k} (n_i - \mu)^2 P(n_i) \qquad (4.7)$$

if n_i is discrete, or

$$\sigma^2 = E[(x - E[x])^2] = \int (x - \mu)^2 P(x) dx = E[x^2] - (E[x])^2 \qquad (4.8)$$

if x is continuous (the integral is over all eligible x).

4.2.2 Binomial Distribution

When a particle passes through a certain readout channel of the detector, usually it provides a "hit" signal or a "miss" signal occasionally. The reason for "miss" may be due to, for example, geometrically dead-regions, dead-time of the readout, or an unexpectedly small gain of the charge from the interaction between the particle and detector material. The probability of "hit" or "miss" is given by the binomial distribution. Let's assume the N particles pass through the detector. The probability of n "hits" and $(N - n)$ "misses" in N trials is given by

$$P(n) = \frac{N!}{n!(N-n)!} p^n (1-p)^{N-n} \qquad (4.9)$$

where p and $(1 - p)$ are the probability of "hit" and "miss" in a single trial, respectively. The sum of the $P(n)$ from $n = 0$ to $n = N$ is represented by the binomial expansion of the $[(1 - p) + p]^N = 1$. This means that the sum of Eq. (4.9) is normalised to be 1. Figure 4.1 show the distributions for $N = 20$ with $p = 0.1, 0.5$, and 0.8.

The expectation value (μ) and variance (σ^2) can be extracted by substituting Eq. (4.9) for Eqs. (4.5) and (4.7):

$$\mu = Np, \qquad (4.10)$$
$$\sigma^2 = Np(1 - p). \qquad (4.11)$$

The detailed calculation can be found in Appendix A.1. In the case of large N than a few 10's and small p such as $p \leq 0.1$, the binomial distribution is approximated by the Poisson distribution of $\mu \sim Np$. In contrast, in the case of large N and moderate p

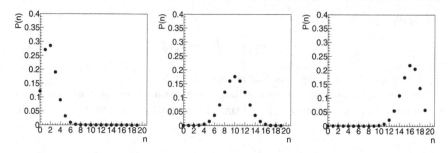

Fig. 4.1 The Binomial distributions represented by Eq. (4.9) for $N=20$ with $p = 0.1, 0.5$, and 0.8

such as $p \sim 0.5$, the binomial distribution is approximated by the normal distribution with the mean and variance expressed by Eqs. (4.10) and (4.11). The Poisson and the normal distributions are explained in the following subsections.

4.2.3   Poisson Distribution

When a theory expects μ events of a certain physics process with some integrated luminosity at a collider experiment, the probability having n events obeys the Poisson distribution, expressed as

$$P(n) = \frac{\mu^n e^{-\mu}}{n!} .$$

(4.12)

Simply put, this distribution can be used in the case where rare events occur. In general, the Poisson distribution describes the probability of n events occurring in a unit interval of time if the events occur with a known average rate μ and independently in the time since the last event. Figure 4.2 show the Poisson distributions for $\mu = 1$, 5, 10, and 20. Using the Maclaurin expansion, i.e. $e^x = \sum_n \frac{x^n}{n!}$, it is shown that the sum of Eq. (4.12) is normalised to 1.

By substituting Eq. (4.12) for Eqs. (4.5) and (4.7), both the expectation value and variance of the Poisson distribution are expressed with only one parameter μ (see Appendix A.2):

$$\mu = mu,$$

(4.13)

$$\sigma^2 = \mu.$$

(4.14)

Then the standard deviation is expressed as $\sigma = \sqrt{\mu}$. It means that you can use the square root of the number of events as a statistical uncertainty in counting experiments. In fact, the mean and the square of the standard deviation of the distributions in Fig. 4.2 are close to μ in Eq. (4.12). But the mean is not exactly the same as the peak position due to the asymmetric shape of the Poisson distribution. If μ become large, for example larger than around 10, the distribution is relatively symmetric and approximated by the normal distribution.

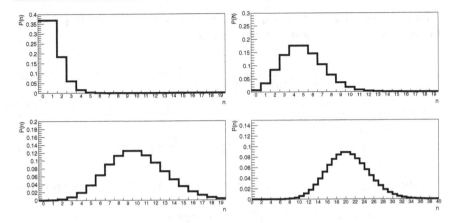

Fig. 4.2 The Poisson distributions represented by Eq. (4.12) for $\mu = 1, 5, 10$, and 20

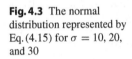

Fig. 4.3 The normal distribution represented by Eq. (4.15) for $\sigma = 10, 20$, and 30

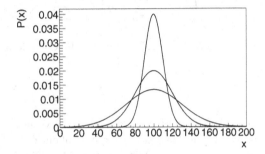

4.2.4 Normal Distribution (Gaussian Distribution)

The normal distribution, which is very often called Gaussian distribution, commonly appears in nature, and used by not only particle physics but also almost everywhere in science. The Gaussian function is symmetric and continuous. It is expressed by two parameters μ and σ,

$$P(x) = \frac{1}{\sqrt{2\pi}\sigma} \exp\left(-\frac{(x-\mu)^2}{2\sigma^2}\right) . \tag{4.15}$$

A coefficient of $\dfrac{1}{\sqrt{2\pi}\sigma}$ is a normalisation factor to ensure that the integral of Eq. (4.15) from $-\infty$ to ∞ is normalised to 1.[1] The Gaussian distributions for $\mu = 100$ and $\sigma = 10, 20$, and 30 are shown in Fig. 4.3.

---

[1] It can be derived using $\displaystyle\int_{-\infty}^{\infty} e^{-a^2 t^2}\, dt = \frac{\sqrt{\pi}}{a}$.

By substituting Eq. (4.15) for Eqs. (4.6) and (4.8), we can find that the parameter μ is the expectation value and σ^2 is the variance:

$$\int_{-\infty}^{\infty} x \cdot \frac{1}{\sqrt{2\pi}\sigma} \exp\left(-\frac{(x-\mu)^2}{2\sigma^2}\right) dx = \mu \qquad (4.16)$$

$$\int_{-\infty}^{\infty} (x-\mu)^2 \cdot \frac{1}{\sqrt{2\pi}\sigma} \exp\left(-\frac{(x-\mu)^2}{2\sigma^2}\right) dx = \sigma^2. \qquad (4.17)$$

For experimental measurements, the values μ and σ are taken from the measured values and the uncertainty.

The integral of the Gaussian distribution in range $\mu \pm \sigma$ is

$$\int_{\mu-\sigma}^{\mu+\sigma} \frac{1}{\sqrt{2\pi}\sigma} \exp\left(-\frac{(x-\mu)^2}{2\sigma^2}\right) dx = \mathrm{erf}\left(\frac{1}{\sqrt{2}}\right) = 0.6827 \qquad (4.18)$$

where the erf(x) is called the error function defined by Eq. (4.19)

$$\mathrm{erf}(x) = \frac{2}{\sqrt{\pi}} \int_0^x e^{-t^2} dx \qquad (4.19)$$

and is shown in Fig. 4.4. Equation (4.18) shows that in the measurement of x, the probability to have $|x - \mu| \le \sigma$ is about 68%. In other words, the probability of $|x - \mu| > \sigma$ is $1 - 0.6827 = 0.3173$ (32%). Several examples of the occurrence having $|x - \mu| > \delta$ are shown in Table 4.2.

Particle physicists use the expression such that a certain measurement has an excess of 5σ from the background-only hypothesis. This means that the number of observed events is larger than the number of events expected from only background events, which is estimated at the 5σ, that is, such a measurement can occur with the probability of 5.73×10^{-7} (for both sides) under the only background environment. This is very rare so we call it "observation" and/or "discovery". In particle physics, we claim "evidence" and "discovery" of something new for 3σ and 5σ excesses, respectively. The full width at half maximum (FWHM) is also often used as the uncertainty rather than σ. This can be easily translated to σ with the equation of FWHM$=2\sqrt{2\ln 2}\sigma = 2.355\sigma$.

4.2.5 Uniform Distribution

The uniform distribution which represents the fixed probability in a certain interval of x is defined as

$$P(x) = \begin{cases} \frac{1}{b-a} & (a \le x \le b) \\ 0 & (\text{otherwise}). \end{cases} \qquad (4.20)$$

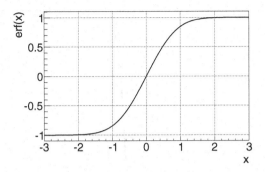

Fig. 4.4 The error function

Table 4.2 The probability outside a certain range expressed in units of σ

δ	Probability (both-side)	Probability (one-side)
1σ	0.3173	0.1587
1.64σ	0.101	0.0505
1.96σ	0.0500	0.0250
2σ	0.0455	0.0228
2.355σ	0.0185	0.00926
3σ	2.70×10^{-3}	1.35×10^{-3}
4σ	6.33×10^{-5}	3.17×10^{-5}
5σ	5.73×10^{-7}	2.87×10^{-7}

We can calculate the expectation value and variance of the uniform distribution by substituting Eq. (4.20) for Eqs. (4.6) and (4.8):

$$\mu = \int x P(x)\mathrm{d}x = \int_a^b \frac{x}{b-a}\mathrm{d}x = \frac{1}{2}(a+b), \tag{4.21}$$

$$\sigma^2 = \int (x-\mu)^2 P(x)\mathrm{d}x = \int_a^b \left\{ x - \frac{1}{2}(a+b) \right\}^2 \frac{1}{b-a}\mathrm{d}x = \frac{1}{12}(b-a)^2. \tag{4.22}$$

An important application of the uniform distribution is position measurements. The position where the particle passes through is determined by position-sensitive sensors. Let's consider the detector with strip-shaped sensors aligned perpendicular to the x axis, which allows you to know the particle position along x. If a certain sensor with a width d, which has a sensitive area from $x = a$ to $x = b$ $(d = b - a)$, provides a hit signal, the expectation value and uncertainty of the position where the particle passes through is estimated to be $\mu = \frac{1}{2}(a+b)$ and $\sigma = \frac{b-a}{\sqrt{12}} = \frac{d}{\sqrt{12}}$, respectively.

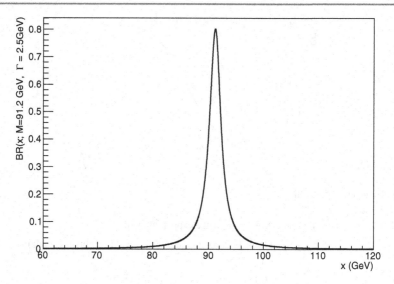

Fig. 4.5 Breit-Wigner distribution for Z boson ($M = 91.2$ GeV, and $\Gamma = 2.5$ GeV)

4.2.6   Breit-Wigner Distribution

The Breit-Wigner distribution is used to express the probability density for the energy of an unstable particle with a mass M and a decay width Γ (and mean lifetime of $\tau = 1/\Gamma$). The Breit-Wigner distribution is defined as

$$\mathrm{BW}(x; M, \Gamma) = \frac{1}{\pi} \frac{\Gamma/2}{(M - x)^2 + (\Gamma/2)^2},$$ (4.23)

and shown in Fig. 4.5. The expectation value and variance of the Breit-Wigner are not well-defined, since the integral of Eqs. (4.6) and (4.8) is divergent. Instead of them, the peak position of M and the FWHM of Γ represent the distribution.

4.2.7   Exponential Distribution

The exponential distribution is used to express the probability density of the existence for the unstable particle with a mean lifetime of τ. The exponential distribution for a continuous variable $0 < x < \infty$ is defined as

$$P(x) = \frac{1}{\tau} e^{-\frac{x}{\tau}},$$ (4.24)

using one parameter τ. The expectation value and variance of x are derived as

$$\mu = \int x P(x) \mathrm{d}x = \frac{1}{\tau} \int_0^\infty x e^{-\frac{x}{\tau}} \mathrm{d}x = \tau,$$ (4.25)

$$\sigma^2 = \int (x - \mu)^2 P(x) dx = \frac{1}{\tau} \int_0^\infty (x - \tau)^2 e^{-\frac{x}{\tau}} dx = \tau^2. \qquad (4.26)$$

4.2.8 χ^2 (Chi-Square) Distribution

In case n observables x_i independently obey the normal distributions $N_i(\mu_i, \sigma_i) = \frac{1}{\sqrt{2\pi}\sigma_i} \exp\left(-\frac{(x_i - \mu_i)^2}{2\sigma_i^2}\right)$, the χ^2 value defined as

$$\chi^2 = \sum_{i=1}^{n} \frac{(x_i - \mu_i)^2}{\sigma_i^2} \qquad (4.27)$$

is used as the test of a hypothesis, which indicates how well the expectation matches with the experimental data. If the hypothesis predicts the nature properly, the $(x_i - \mu_i)^2$ is expected to be the variance of the experiment, i.e. σ^2. Thus, χ^2/n is expected to be 1.

If χ^2/n shows significant deviation from unity, either the hypothesis or the estimation of the σ's is wrong. The probability density function of this χ^2 distribution with n degrees of freedom (dof) can be written as

$$f(z; n) = \frac{z^{n/2-1} e^{-z/2}}{2^{n/2} \Gamma(n/2)} \qquad (z > 0), \qquad (4.28)$$

where Γ is the Γ function. Figure 4.6 shows the χ^2 distribution $f(\chi^2, n)$ for dof $n = 1$ to 5. For large n, the probability density function of this χ^2 distribution approaches the normal distribution with a mean and variance of $\mu = n$, $\sigma^2 = 2n$, respectively.

4.3   Error Propagation

As the cross section is determined by the values of the five parameters in Eq (4.1), a physical quantity is often derived from several parameters which is determined by measurements. Naturally, the uncertainty of parameters carries over into a physical quantity. Let's assume a physical quantity u depends on ith parameters x_i. Namely, the u can be written as a function of x_i ($i = 1, 2, \ldots, n$): $u = f(x_1, x_2, \ldots, x_n) = f(x)$. The expectation values and uncertainties of x are known as $\mu = (\mu_1, \mu_2, \ldots, \mu_n)$ and $\sigma_i = (\sigma_1, \sigma_2, \ldots, \sigma_n)$, respectively. A first-order expansion of the function $f(x)$ around the expectation value μ can be written as

$$f(x) \approx f(\mu) + \sum_{i=1}^{n} \left. \frac{\partial f(x)}{\partial x_i} \right|_{x=\mu} (x_i - \mu_i). \qquad (4.29)$$

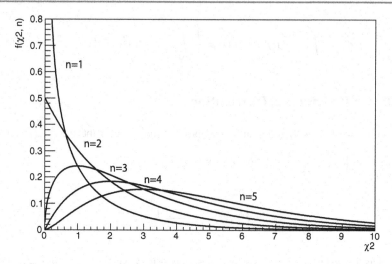

Fig. 4.6 χ^2 distributions. See the main text for details

Because $E[x_i - \mu_i] = 0$, the expectation value of u is represented at the first order to be

$$E[f(\boldsymbol{x})] \approx f(\boldsymbol{\mu}), \tag{4.30}$$

and the expectation value of u^2 is

$$
\begin{aligned}
E[f^2(\boldsymbol{x})] &\approx f^2(\boldsymbol{\mu}) \\
&+ E\left[\left(\sum_{i=1}^{n} \frac{\partial f(\boldsymbol{x})}{\partial x_i}\bigg|_{\boldsymbol{x}=\boldsymbol{\mu}} (x_i - \mu_i)\right)\left(\sum_{j=1}^{n} \frac{\partial f(\boldsymbol{x})}{\partial x_j}\bigg|_{\boldsymbol{x}=\boldsymbol{\mu}} (x_j - \mu_j)\right)\right] \\
&= f^2(\boldsymbol{\mu}) + \sum_{i=1}^{n} \left(\frac{\partial f(\boldsymbol{x})}{\partial x_i}\bigg|_{\boldsymbol{x}=\boldsymbol{\mu}}\right)^2 E[(x_i - \mu_i)^2] \\
&+ \sum_{i \neq j}^{n} \frac{\partial f(\boldsymbol{x})}{\partial x_i}\bigg|_{\boldsymbol{x}=\boldsymbol{\mu}} \frac{\partial f(\boldsymbol{x})}{\partial x_j}\bigg|_{\boldsymbol{x}=\boldsymbol{\mu}} E[(x_i - \mu_i)(x_j - \mu_j)].
\end{aligned} \tag{4.31}
$$

In case x_i and x_j are not correlated, the third term of Eq. (4.31) is 0. Because $E[(x_i - \mu_i)^2] = \sigma_i^2$, the variance of the u can be calculated to be

$$\sigma_u^2 = E[f^2(\boldsymbol{x})] - (E[f(\boldsymbol{x})])^2 \approx \sum_{i=1}^{n} \left(\frac{\partial f(\boldsymbol{x})}{\partial x_i}\bigg|_{\boldsymbol{x}=\boldsymbol{\mu}}\right)^2 \sigma_i^2. \tag{4.32}$$

Suppose in the cross-section measurement, N_{obs}, N_{bkgd}, L, A, and ε can be measured independently. In this case, (the square of) the uncertainty of the cross section ($\sigma_{\sigma_{\text{phys}}}^2$)

can be expressed as

$$
\sigma^2_{\sigma_{\text{phys}}} = \left(\frac{\partial \sigma_{\text{phys}}}{\partial N_{\text{obs}}} \right)^2 \cdot \sigma^2_{N_{\text{obs}}} + \left(\frac{\partial \sigma_{\text{phys}}}{\partial N_{\text{bkgd}}} \right)^2 \cdot \sigma^2_{N_{\text{bkgd}}} + \left(\frac{\partial \sigma_{\text{phys}}}{\partial L} \right)^2 \cdot \sigma^2_L
$$
$$
+ \left(\frac{\partial \sigma_{\text{phys}}}{\partial A} \right)^2 \cdot \sigma^2_A + \left(\frac{\partial \sigma_{\text{phys}}}{\partial \varepsilon} \right)^2 \cdot \sigma^2_\varepsilon. \tag{4.33}
$$

Imagine you measure the efficiency of the particle detection for a detector. When N particles passed through the detector, the detector gives "hit" signal n_1 times and "miss" signal n_2 times ($N = n_1 + n_2$). In this case, the detection efficiency is given by $\varepsilon = \dfrac{n_1}{n_1 + n_2}$. If the n_1 and n_2 are large enough to consider that they are not correlated and the their uncertainties are estimated as $\sqrt{n_1}$ and $\sqrt{n_2}$, respectively, you can show the uncertainty of efficiency, σ_ε to be $\sqrt{\dfrac{\varepsilon(1 - \varepsilon)}{N}}$. Similar discussion can be done for the asymmetry $A = \dfrac{n_1 - n_2}{n_1 + n_2}$, instead of ε. Show by yourself the uncertainty of the asymmetry, σ_A.

4.4 Maximum Likelihood Method

Although we can never know the true values of physical quantities, we can estimate them from a set of the measurements. Consider that we made n independent measurements and obtained n measured quantities $x = (x_1, x_2, ..., x_n)$. Suppose that the measured quantities x distributes a probability density function $f(x_i; \theta)$ ($i = 1, 2, ..., n$), where $\theta = (\theta_1, \theta_2, ..., \theta_m)$ are unknown physical quantities. The likelihood function $L(x; \theta)$, which is regarded as the probability to have a set of measurements of $(x_1, x_2, ..., x_n)$, is defined as

$$
L(x; \theta) = f(x_1; \theta) f(x_2; \theta) \cdots f(x_n; \theta) = \prod_{i=1}^{n} f(x_i; \theta). \tag{4.34}
$$

If the hypothesis constructing the probability density function $f(x; \theta)$ and parameter values θ are correct, one expects that the L gives maximum. To estimate the most probable values of θ, the maximum likelihood estimators for θ are defined as the values which maximise the likelihood function. As long as the likelihood function is a differentiable function of the parameters θ, and the maximum is not at the physical boundary, the estimators are given by solving the simultaneous equations

$$
\frac{\partial L}{\partial \theta_i} = 0, \quad \text{or} \quad \frac{\partial \ln L}{\partial \theta_i} = 0, \quad i = 1, 2, ..., m. \tag{4.35}
$$

Because of the characteristics of the logarithm, maximum log-likelihood estimators, which are equivalent to the maximum likelihood, are often used. To distinguish the true values of physical quantities ($\boldsymbol{\theta} = (\theta_1, \theta_2, ...\theta_m)$) from their estimators, the parameters satisfied with Eq. (4.35) are written as $\hat{\boldsymbol{\theta}} = (\hat{\theta}_1, \hat{\theta}_2, ...\hat{\theta}_m)$.

As an example, let's consider that a variable x, which obeys a Gaussian distribution with unknown μ and σ^2, is measured n times. The log-likelihood function is

$$\ln L(\mu, \sigma^2) = \sum_{i=1}^{n} \ln f(x_i; \mu, \sigma^2) = \sum_{i=1}^{n} \ln \frac{1}{\sqrt{2\pi}\sigma} \exp\left(-\frac{(x_i - \mu)^2}{2\sigma^2}\right)$$

$$= \sum_{i=1}^{n}\left(-\ln\sqrt{2\pi} - \frac{1}{2}\ln\sigma^2 - \frac{(x_i - \mu)^2}{\sigma^2}\right). \tag{4.36}$$

By solving $\dfrac{\partial \ln L}{\partial \mu} = 0$, $\hat{\mu}$ is obtained as

$$\hat{\mu} = \frac{1}{n}\sum_{i=1}^{n} x_i. \tag{4.37}$$

The expectation value of $\hat{\mu}$ is an unbiased estimator for μ:

$$E(\hat{\mu}) = \mu. \tag{4.38}$$

This calculation can be found in Appendix A.3. Similarly, solving $\dfrac{\partial \ln L}{\partial \sigma^2} = 0$ gives $\hat{\sigma}^2$

$$\hat{\sigma}^2 = \frac{1}{n}\sum_{i=1}^{n}(x_i - \mu)^2 = \frac{1}{n}\sum_{i=1}^{n}(x_i - \hat{\mu})^2. \tag{4.39}$$

Because μ is an unknown parameter, $\hat{\mu}$ is actually used to estimate σ. Computing the expectation value of σ^2, it gives

$$E[\hat{\sigma}^2] = \frac{n-1}{n}\sigma^2, \tag{4.40}$$

which means that the estimator $\hat{\sigma}^2$ is biased, because using $\hat{\mu}$ instead of μ reduces the number of dof by 1. Instead of $\hat{\sigma}^2$,

$$s^2 = \frac{n}{n-1}\hat{\sigma}^2 = \frac{1}{n-1}\sum_{i=1}^{n}(x_i - \mu)^2 = \frac{1}{n}\sum_{i=1}^{n}(x_i - \hat{\mu})^2 \tag{4.41}$$

may be used as a more correct estimator, but the difference between them is ignored when n is large enough.

4.5 Least Squares Method

Suppose $y = f(x; \theta_1, \theta_2, \ldots, \theta_m)$ is a function of x and you want to determine m parameters of $\boldsymbol{\theta} = (\theta_1, \theta_2, \ldots, \theta_m)$ by measuring parameters y_i at the n points of x_i ($i = 1, 2, \ldots, n$). When the uncertainties of measurements y_i are given by σ_i, the parameters can be estimated by finding the values of the parameter $\boldsymbol{\theta}$ that minimise the following quantity:

$$\chi^2 = \sum_{i=1}^{n} \frac{(y_i - f(x_i; \boldsymbol{\theta}))^2}{\sigma_i^2}. \qquad (4.42)$$

This method is called the least squares method. This method is equivalent to the maximum likelihood method described in Sect. 4.4. In fact, when $f(x_i; \mu, \sigma^2)$ obeys a Gaussian distribution, χ^2 is identical to $-2 \ln L(\mu, \sigma^2)$.

4.6 Statistical Figure of Merit

When we discuss the statistical significance of observed events, the following figure of merit is often used,

$$\frac{N_{\text{signal}}}{\sqrt{N_{\text{obs}}}} = \frac{N_{\text{signal}}}{\sqrt{N_{\text{signal}} + N_{\text{background}}}}. \qquad (4.43)$$

Because the $\sigma = \sqrt{N_{\text{obs}}}$ shows the statistical uncertainty of total number of observed events, the figure of merit above is the indicator to show how significant we have the signal over the background in units of σ. For example, suppose 10000 events are observed and 9500 events are expected as background events after a certain event selection, the figure of merit is $(10000 - 9500)/(\sqrt{10000}) = 5\sigma$. If the $N_{background}$ is much larger than N_{signal}, one can use

$$\frac{N_{\text{signal}}}{\sqrt{N_{\text{background}}}}. \qquad (4.44)$$

The higher the statistical figure of merit, the more sensitive the measurement can be expected. Note that in this discussion, only statistical uncertainty is taken into account. If you need to consider systematic uncertainties, the figure of merit becomes more complicated.

4.7 Hypothesis Test

A hypothesis test is a method to describe how well the data agree or disagree with a given hypothesis. The hypothesis under consideration is called the null hypothesis H_0. This hypothesis H_0 is compared with a so-called alternative or test hypothesis H_1 in order to quantify the compatibility of H_0. In practice, H_1 is a hypothesis we

Fig. 4.7 Probability density distributions for H_0 (left) and H_1 (right) hypotheses. Given a threshold on the number of events ("thres"), the regions for the *significance level* of α and the *power* $1 - \beta$ are shown

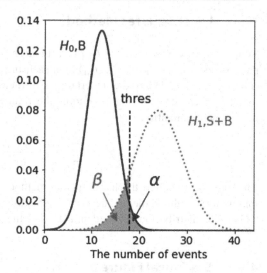

would like to see, for example, the presence of a new particle. In other words, H_0 is a hypothesis we would like to reject.

4.7.1 Discovery and Exclusion

According to the Neyman-Pearson lemma, the likelihood ratio $L(H_0)/L(H_1)$ is the optimal discriminator for the hypothesis test H_0 versus H_1 such that we often use the likelihood ratio as a *test statistic t*. However, in this section we use the number of events we select as a test statistic assuming they follow Gaussian distributions (Eq. (4.15)) to understand p-value, etc. intuitively, where we discuss *discovery* and *exclusion* using the number of selected events. Suppose that a Gaussian distribution $N_B(x)$ is for background-only (the SM) and the other one $N_{S+B}(x)$ is for signal+background (the signal is a new physics beyond the SM). Figure 4.7 shows these two Gaussian distributions. Since the x-axis is the number of events, the $N_{S+B}(x)$ distribution is present in the right side of $N_B(x)$. Here, we assume that the background-only model is the null hypothesis H_0 and the signal+background is for H_1. For a hypothesis test, we determine a threshold x^{thres} to define a *significance level* α. In this case, the α is defined as

$$\alpha = \int_{x^{\text{thres}}}^{\infty} N_{H_0:B}(x)dx,$$

which is shown in Fig. 4.7. Then, if H_0 is false and H_1 is true, the probability to reject H_0 correctly is called a *power* $1 - \beta$ where the β is defined as

$$\beta = \int_{-\infty}^{x^{\text{thres}}} N_{H_1:S+B}(x)dx,$$

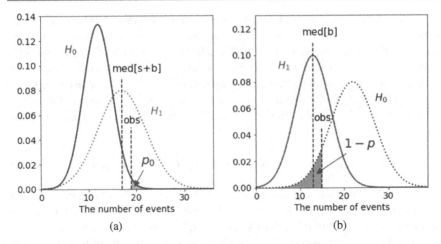

Fig. 4.8 a p_0 for discovery and **b** $1 - p$ for exclusion. They are evaluated for the number of observed events ("obs"). In case of MC studies, the number of observed events is replaced with the median of signal+background and background-only for **(a)** and **(b)**, respectively

which is also shown in Fig. 4.7. These α and β also correspond to *type I error (false positive)* and *type II error (false negative)*, respectively. The former is the probability to reject H_0 wrongly and the latter is to reject H_1 wrongly. Then, we assume that we obtain the number of events x^{obs} from the data. We define a *p*-value p, which is a probability to show the compatibility with H_0:

$$p = \int_{x^{\mathrm{obs}}}^{\infty} N_{H_0:B}(x)dx.$$

When the value of p is smaller than α, we can say that H_0 is rejected by the *significance level* of α.

Figure 4.8a shows an example of the discovery, where H_0 is the background-only and H_1 is the signal+background. We use the *p*-value p_0 under H_0 to claim the discovery of a new particle.[2] Conventionally, if p_0 is smaller than 2.87×10^{-7}, what we observe is very rare under H_0 such that we consider that H_0 is rejected. The *p*-value can be transformed to *z*-value, which is defined using a standard Gaussian distribution as

$$p = \int_z^{\infty} \frac{1}{\sqrt{2\pi}} e^{-\frac{x^2}{2}} dx.$$

For $p = 2.87 \times 10^{-7}$, *z*-value corresponds to 5σ, which is shown in Table 4.2. When we investigate new physics models using MC samples (MC studies), the observed x^{obs} is replaced with the median of the $N_{H_1:S+B}(x)$ distribution. To claim the so-called "evidence" instead of "discovery", we conventionally use $p = 1.35 \times 10^{-3}$ (3σ).

---

[2] The suffix of 0 is often used for the background-only.

Figure 4.8b shows an example of the exclusion of a model, where H_0 is the signal+background and H_1 is the background-only. When the value of $(1 - p)$ under H_0 is smaller than 0.05, H_0 is rejected. We call it "95% exclusion."[3] In case of MC studies, the x^{obs} is replaced with the median of the $N_{H_1:B}(x)$ distribution. The $(1 - p)$ is denoted as CL_{s+b} in high-energy experiments since the $(1 - p)$ value corresponds to compatibility with the signal+background hypothesis. Furthermore, in the LHC experiments, a CL_s-based exclusion is often used instead of CL_{s+b}. CL_s is defined as

$$CL_s = \frac{CL_{s+b}}{CL_b} = \int_{-\infty}^{x^{\text{obs}}} N_{H_0:S+B}(x)dx / \int_{-\infty}^{x^{\text{obs}}} N_{H_1:B}(x)dx. \qquad (4.45)$$

In case of MC studies, the denominator is 0.5 because x^{obs} is the median of the $N_{H_1:B}(x)$ distribution; CL_s is $2CL_{s+b}$ so that the 95% exclusion using CL_s corresponds to the $(1 - p)$ of 0.025 for CL_{s+b}. The CL_s is not a probability but in order to avoid incorrect exclusions, which could be possible when the expected signal is small, the LHC experiments often use it to claim exclusions.

4.7.2   Profile Likelihood Fit

Suppose we count the number of observed events n after applying our event selection. This parameter of n follows a Poisson distribution with an expectation value of $\mu s + b$, where s is the expected value from a signal model, and b is the expected value from background processes. The likelihood function can be defined as

$$L(\mu, s, b) = \frac{(\mu s + b)^n}{n!} e^{-(\mu s + b)}.$$

The parameter of μ is a scale factor of the signal and is called a *signal strength*. Given s and b, we can extract a μ value from a fit to data, which gives the value of n, using the maximum likelihood technique explained in Sect. 4.4. The value of μ be around unity if the data follows the assumed signal model, while it is close to zero if the data follows the SM, that is, the data contain background only.

We modify this likelihood function by adding more terms. Since the s corresponds to $L\sigma_{\text{phys}} A\varepsilon$ of Eq. (4.1), we can consider systematic uncertainties from these parameters $(L, \sigma_{\text{phys}}, A, \varepsilon)$. For example, the uncertainty on the integrated luminosity, the scale uncertainties (factorisation and renormalisation scales) on the σ_{phys}, the uncertainties from jet energy scale, etc. on the ε and so on can be systematic uncertainties for μ. We often use Gaussian terms[4] to constrain the signal term and also other terms in b. Furthermore, in most of data analyses, we estimate the background b in

---

[3] In some experimental results, 90% is also used instead of 95%. For 90%, the value of $(1 - p)$ is set to be 0.1.

[4] A log-normal term, etc. can be used instead of a Gaussian term, for example, if we require a positive definition.

the signal region, which is defined by our (signal) event selection, by using a so-called control or sideband regions in data with the help of MC samples. In this case, the b in the signal region can be described with $\eta_{\text{tf}}(\alpha_{\text{tf}})b$, where b is the number of events in the control region, and η_{tf} is a scale (transfer) factor from the control region to the signal region. The η_{tf} is obtained from both data and MC samples so that some additional constraints (α_{tf}) are possible. In the end, one of examples of a final likelihood function can be written as

$$L(\mu, \boldsymbol{\theta}) = Pois(n|\mu\eta_s(\alpha_s)s + \eta_{\text{tf}}(\alpha_{\text{tf}})b) \cdot N(\alpha_s|0, 1)\cdot$$
$$Pois(m|\eta_b(\alpha_b)b) \cdot N(\alpha_b|0, 1)\cdot$$
$$N(\alpha_{\text{tf}}|0, 1),$$

where $\boldsymbol{\theta} = (b, \alpha_s, \alpha_b, \alpha_{\text{tf}})$, $\eta_i(\alpha) = \mu_i + \sigma_i\alpha$, $Pois(n|\mu) = \mu^n e^{-\mu}/n!$, and $N(x|\mu, \sigma) = 1/(\sqrt{2\pi}\sigma) \cdot \exp(-(x - \mu)^2/(2\sigma^2))$. The m is the number of observed events in the control region. The $\boldsymbol{\theta}$ is called a set of *nuisance parameters*, which are determined by the likelihood fit with the μ. The η_i is a scale parameter for signal, background, and so on. The α_i is a parameter to adjust the η_i through a Gaussian constraint. Parameters μ_i and σ_i for η_i describe centre values and their uncertainties and are evaluated from other studies before the likelihood fit. If the α_i value is 0, the value of η_i becomes μ_i. If not, the value of η_i is varied from its centre value. Practically, μ_i is close to 1. Then, the effect on the signal strength μ from each constraint is determined in the maximum likelihood (ML) fit. It means that the systematic uncertainties on the μ from each constraint term are simultaneously determined with the μ value itself. We call this procedure a "profiled" fit. When the pre-studies on μ_i and σ_i are proper, the values of α_i are expected to be close to 0 ± 1. For example, if the error of α_i is smaller than 1 (say 0.3 or 0.4), it means that the value of σ_i given from the pre-studies is tightly constrained from data used in the ML fit, for example, data of control regions. If this is not expected, some additional studies might be required to understand such small values.

4.7.3 Profile Likelihood Ratio

We introduce the following likelihood ratio as a test statistic t_μ:

$$t_\mu = -2\ln\lambda(\mu),$$

$$\lambda(\mu) = \frac{L(\mu, \hat{\hat{\boldsymbol{\theta}}})}{L(\hat{\mu}, \hat{\boldsymbol{\theta}})},$$

where the denominator of $\lambda(\mu)$ is maximised for both μ and $\boldsymbol{\theta}$ (an unconditional ML fit) but the numerator is maximised for $\boldsymbol{\theta}$ with respect to a specified μ value (a conditional ML fit). Since the denominator corresponds to the best fit to data, the value of $\lambda(\mu)$ is $0 < \lambda(\mu) \leq 1$ so that the value of t_μ is 0 or positive. When the numerator with a specified μ value follows the data, the t_μ can be small, if not, the t_μ becomes large. The p-value is defined as

$$p_\mu = \int_{t_{\mu,\text{obs}}}^{\infty} f(t_\mu|\mu')dt_\mu,$$

where $t_{\mu,\text{obs}}$ is the value of the observed t_μ and $f(t_\mu|\mu')$ is the probability density distribution of t_μ under the assumption of the signal strength μ'. The advantage of the use of this test statistic is that the distribution of t_μ follows a χ^2 distribution of one degree-of-freedom: $f(t_\mu|\mu) \sim \chi^2_{\text{dof}=1}(t_\mu)$, so that we can evaluate p-value without toy Monte Carlo.[5] We explain the overall idea of the discovery and exclusion using this test statistic below but the technical detail of the hypothesis test using this test statistic can be found in Ref. [1].

In high-energy experiments, we search for a new signal particle by checking an excess over the expected events of a background-only assumption. The signal existence corresponds to $\mu > 0$. For this case, an alternative test statistic $\tilde{t}_\mu = -2 \ln \tilde{\lambda}(\mu)$ is introduced,

$$\tilde{\lambda}(\mu) = \begin{cases} \frac{L(\mu,\hat{\hat{\theta}}(\mu))}{L(\hat{\mu},\hat{\theta}(\hat{\mu}))} & (\hat{\mu} \geq 0) \\ \frac{L(\mu,\hat{\hat{\theta}}(\mu))}{L(0,\hat{\hat{\theta}}(0))} & (\hat{\mu} < 0), \end{cases} \tag{4.46}$$

where the best-fit μ value with a deficit ($\hat{\mu} < 0$) is replaced with $\mu = 0$.

4.7.3.1 Discovery

We test $\mu = 0$, that is, we reject the null hypothesis H_0 of $\mu = 0$ (background-only). We use a special notation $q_0 = \tilde{t}_0$ for this case. From Eq. (4.46), we use

$$q_0 = \begin{cases} -2 \ln \lambda(0) = -2 \ln \frac{L(0,\hat{\hat{\theta}}(0))}{L(\hat{\mu},\hat{\theta}(\hat{\mu}))} & (\hat{\mu} \geq 0) \\ 0 & (\hat{\mu} < 0). \end{cases}$$

We get a single value of q_0 from data, q_0^{obs}, and evaluate p-value p_0 using

$$p_0 = \int_{q_0^{\text{obs}}}^{\infty} f(q_0|0)dq_0,$$

where the $f(q_0|0)$ is a distribution of q_0 made under the assumption of $\mu = 0$. Figure 4.9a shows distributions of q_0 for the assumption of $\mu = 0$ and 1: $f(q_0|0)$

---

[5] For $f(t_\mu|\mu')$, we can use a noncentral χ^2 distribution of one degree-of-freedom.

and $f(q_0|1)$. Once we obtain a distribution $f(q_0|0)$, we can evaluate p_0 using q_0^{obs}. Then, when the p_0 value is smaller than 2.87×10^{-7}, we can claim discovery.[6]

We can approximate the $f(q_0|0)$ distribution as follows:

$$f(q_0|0) = \frac{1}{2}\delta(q_0) + \frac{1}{2}\frac{1}{\sqrt{2\pi}}\frac{1}{\sqrt{q_0}}e^{-q_0/2}.$$

By using this equation, a z-value Z can be obtained as

$$Z = \sqrt{q_0^{\text{obs}}}.$$

The 5σ discovery corresponds to $q_0^{\text{obs}} = 25$. In Fig. 4.9a, the value of q_0^{obs} is 23, which is just an example, so that we cannot say the "discovery." In case of MC studies, as shown in Fig. 4.9a, we can use the median of the $f(q_0|1)$ distribution as q_0^{obs}. In this case, $q^{\text{med}}[f(q_0|1)]$ is smaller than 25 so that we cannot claim the "discovery"[7] by a physics model of $s = 20$.

4.7.3.2 Exclusion or Upper Limit

We test $\mu(\neq 0)$, that is, we reject the null hypothesis H_0 of the signal+background model. When a specified μ value is equal to or smaller than the $\hat{\mu}$ of the unconditional ML fit, we consider that q_μ is 0. It means that the exclusion of models is performed for only μ values which are larger than the observed best-fit μ. We define q_μ as

$$q_\mu = \begin{cases} -2\ln\lambda(\mu) & (\hat{\mu} \le \mu) \\ 0 & (\hat{\mu} > \mu). \end{cases}$$

We evaluate p-value p_μ using

$$p_\mu = \int_{q_\mu^{\text{obs}}}^{\infty} f(q_\mu|\mu')dq_\mu,$$

where the $f(q_\mu|\mu')$ is a distribution of q_μ made under the assumption of μ'. Figure 4.9b shows distributions of $q_{\mu=1}$ (simply q_1) for the assumption of $\mu = 0$ and 1. When the p_μ value is smaller than 0.05, we can claim 95% CL_{s+b} exclusion. This corresponds to $q_0^{\text{obs}} > 2.69 (= 1.64^2)$. In Fig. 4.9b, the observed q_1 (23

---

[6] In some special cases, we may take into account so-called *look-elsewhere effect*. In this case, the standard p_0 value is called a local p_0 and the p_0 after the look-elsewhere effect is called a global p_0. In case of the Higgs search in the LHC experiments (see Sect. 8.1.2), this effect was taken into account because the Higgs mass is unknown and we searched for a Higgs signal in a certain mass range which is much wider than resolutions (Higgs natural width $\otimes$ detector), where we expect some statistical fluctuations even if we would have a true Higgs. The practical method is discussed in Ref. [2].

[7] The value of $q^{\text{med}}[f(q_0|1)]$ is larger than 9 so that we may say the "evidence." For the "discovery", we may need $s \simeq 36$ in this example.

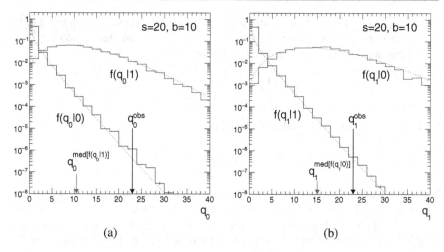

(a)   (b)

Fig. 4.9 **a** $f(q_0|0)$ with $f(q_0|1)$ for discovery **b** $f(q_1|1)$ with $f(q_1|0)$ for exclusion (upper limit). $f(*|0)$ and $f(*|1)$ show distributions for background-only and signal+background events, respectively. A likelihood function of $L(\mu, \theta = b) = \frac{(\mu s+b)^n}{n!}e^{-(\mu s+b)} \cdot \frac{b^m}{m!}e^{-b}$ is used, where variables of n and m are the number of events observed in the signal and control regions and $s = 20$ and $b = 10$ are used in this example; b in the signal region is estimated from the value of b in the control region. Dashed curves are central (for blue) and noncentral (for red) χ^2 distributions of one degree-of-freedom. For the noncentral cases, so-called Asimov data, which is defined as data produced with the expectation values of inputs (s, b, and μ), is used to evaluate a width required in an approximate formula of $f(q_\mu|\mu')$ [1]

as an example) can claim the "95% CL_{s+b} exclusion", where a model of $s = 20$ cannot be explained. Practically, we need a scan of μ values to find a μ value having $p_\mu = 0.05$. This corresponds to $\mu \sim 0.4$ in case of Fig. 4.9b. For 95% CL_s exclusion[8], we need a distribution $f(q_\mu|0)$ to evaluate $CL_b = \int_{q_\mu^{obs}}^{\infty} f(q_\mu|0)dq_\mu$. In case of MC studies, as shown in Fig. 4.9b, we can use the median of the $f(q_\mu|0)$ distribution as q_μ^{obs} and $CL_b = 0.5$. For 95% CL_s exclusion of MC studies, we can use $q_\mu^{obs} > 3.84 (= 1.96^2)$.

For the case where we consider models with $\mu \geq 0$, we can define and use an alternative test statistic $\tilde{q}_\mu$:

$$\tilde{q}_\mu = \begin{cases} -2\ln\frac{L(\mu,\hat{\hat{\theta}}(\mu))}{L(0,\hat{\hat{\theta}}(0))} & (\hat{\mu} < 0) \\ -2\ln\frac{L(\mu,\hat{\hat{\theta}}(\mu))}{L(\hat{\mu},\hat{\theta}(\hat{\mu}))} & (\mu \geq \hat{\mu} \geq 0) \\ 0 & (\hat{\mu} > \mu). \end{cases}$$

The procedure similar to the case of q_μ can be applied [1].

---

[8] Note that the integration range is different to the case of Eq. (4.45).

References

1. Cowan, G., Cranmer, K., Gross, E., Vitells, O.: Eur. Phys. J. C **71**, 1554 (2011). [erratum: Eur. Phys. J. C **73**, 2501 (2013)]. https://doi.org/10.1140/epjc/s10052-011-1554-0, arXiv:1007.1727 [physics.data-an]
2. Gross, E., Vitells, O.: Eur. Phys. J. C **70**, 525–530 (2010). https://doi.org/10.1140/epjc/s10052-010-1470-8, arXiv:1005.1891 [physics.data-an]

Detector Calibration

5

The information needed when analysing data, for example, the cross-section measurements, is the 4-momentum vectors of particles in interest, and possibly the knowledge about the species of the particles. For this purpose, the detector for the high energy physics is usually designed so that it allows us to measure the momentum or energy, and information for the particle identification, such as velocity. However, the recorded data as they are do not tell us anything. They are just a bunch of digits which are not energies or positions of particles if they are not properly translated into meaningful physical variables.

This chapter describes the procedure to retrieve meaningful information that is needed in physics analyses from raw data. This process is called calibration, and is one of the most important processes in the whole flow of the high energy physics experiments.

5.1   From Raw Data to Meaningful Information

Let's first imagine how data is generated and recorded. As an example, suppose we take data of electromagnetic calorimeter consisting of the sandwich structure of lead and scintillator. A photon hitting the calorimeter generates position-electron pair by photon conversion mainly in the lead plates. The positron or electron ionises the scintillator, resulting in scintillation light. This scintillation light is detected by some photo-sensors such as the photomultiplier tube (PMT). The electrical signal from the PMT is then converted by an analog-to-digital converter (ADC) and recorded as the series of a bunch of digits. Ideally, the light yield of the scintillator is linear to the energy deposit by the electron or position, and also the PMT output is linear to the scintillation light. With the assumption of the linearity of the light yield and the PMT response, one can measure the energy deposited by the electron and positron,

© The Author(s) 2022

K. Hanagaki et al., *Experimental Techniques in Modern High-Energy Physics*,
Lecture Notes in Physics 1001, https://doi.org/10.1007/978-4-431-56931-2_5

or ultimately the incident photon energy from the ADC counts in principle. However, let's recall that what we have here is just ADC counts which are solely digits. They represent the energy, but do not mean energy yet without the proper conversion to energy. This important procedure to convert the ADC counts to energy is called a "detector calibration", in short "calibration".

In the above example, we discussed the concept of energy calibration of the calorimeter. This concept is very common for all detectors, whatever they measure. Sometimes, we measure the time interval of some detector signals, in which the data is recorded by time-to-digital converter (TDC). In this case, the calibration from TDC counts to time is needed. Sometimes, we measure the location of a charged particle hitting the position-sensitive sensor. In this case, the hit information has to be interpreted as the position information. This is also considered as a calibration in a sense. In the following sections, we discuss some concrete procedures of the detector calibrations.

5.2   Detector Alignment

The tracking device usually consists of many finely segmented channels. Assuming we know the location of each channel, we can measure the position of charged particles by sensing the signal from each channel. This means that the accurate and precise knowledge of the location of the tracking device, or more precisely the position of each channel, including the angles hence six degrees of freedom, is crucial to measure the hit position of the particle at the detector. The procedure to retrieve the position of the tracking device or each channel is called "alignment". Not only the tracking devices but also any other detectors consisting of multi-channel detectors need to be aligned as well.

The alignment procedure can be divided into two steps. The first one is the mechanical measurement or survey of the detector component. In the survey, the location of the large structure of the detector is measured, which has to be carried out usually before starting data taking or just after installing the detector. Since the detector element of the large structure is assembled from the small components of sensors, etc. with some precision which is specified in each experiment, the position of each channel is considered to be known with the precision of the assembly (and the survey) once the survey is performed.

The second step is the alignment using charged particles. The idea is to use such charged particles as a probe. Suppose we have tracking device composed of five layers of silicon strip detectors, and we would like to align the sensors of a particular layer. In this case, a special tracking algorithm, which does not use the information on the layer that will be aligned, should be prepared and the particle trajectory reconstructed. Then this track is extrapolated to the layer under alignment to get the so-called residual, the difference in position between the extrapolated probe track and the hit on the layer. For this reason, a higher momentum track is preferred to minimise the extrapolation uncertainty due to the multiple scattering. Here, when we say a "hit", it is based on the hypothesis or the prior knowledge of the location of the silicon strip sensor. If this prior knowledge is wrong, the residual cannot be zero. In other words, the sensor position that gives the zero residual is likely to be the true position. Based on this general idea, the layer under the alignment is aligned by adjusting the location of the sensors so that the residual distributions have a peak close to zero and the width to be narrow.

In the actual application, a slightly different approach is taken although the basic idea is the same as we have just explained. Because there are many layers and millions of channels in the tracking detectors of the modern collider experiment, it is time-consuming and complicated to prepare such tracking algorithm that doesn't use the information from the specific layers or channels. Instead of having such a special algorithm, it is more common to use the normal track reconstruction algorithm with looser quality requirements to minimise the bias arising from the usage of hit information from the layer or channel under alignment. For each probe track, again the residual is measured. But not only for a single layer or channel but also for all the layers or channels in the detector under alignment, the residual is computed. Then the sum of residuals from many layers is calculated for each track. The detector is aligned so that the total residual is minimised. This is almost equivalent to the χ^2 minimisation where the positions of the sensors are fitted.

So far in this section, the basic concept of the alignment was given, where we discussed the alignment of the single detector. But the collider detector, for example, is a more complex and larger object consisting of several types of detectors. Further in the actual application, the alignment is performed in several steps. Again, using the silicon strip tracker in the ATLAS experiment as an example, the first level is to align the whole tracker relative to the other detector system. This means that not each layer nor single sensor is individually aligned. Instead, the whole support structure holding the sensors or modules is aligned as a single object. Then as the second-level alignment, each layer is aligned, i.e. each layer can be moved independently. Finally as the third level, the individual module or sensor within each layer is aligned. In this way, the failure of the χ^2 fitting due to the possible large deviation of the initial value from the actual position can be avoided. In addition, the step-by-step approach allows saving the computing time of the χ^2 fitting.

Figure 5.1 shows the residual distributions for the ATLAS silicon pixel detector, where the first alignment was carried out by using cosmic rays and then proton-proton collision data for more statistics. You can see that the width becomes narrower by using the collision data, indicating the improvement of the alignment. Note that an old result, which was obtained at the very beginning of the experiment, is intentionally presented here for the illustration purpose. Currently, the width of the residual distribution is close to that for the simulation result where all the detector positions are perfectly known.

5.3 Momentum Scale Calibration of Magnetic Spectrometer

We first explain the concept of measurements of charged particle momentum. Then the calibration of the momentum scale is discussed.

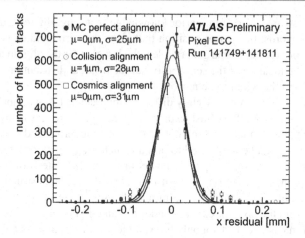

Fig. 5.1 The residual distribution for the silicon pixel detector of the ATLAS experiment. Reprinted under the Terms of Use from [1] ATLAS Experiment © 2022 CERN. All rights reserved. Red (black) shows the residual using proton-proton collision (cosmic ray) data. Blue shows the prediction by simulation. Note this plot is intentionally selected for the illustration purpose of the effect of the alignment, not showing the current precision

Fig. 5.2 Sagitta measurement. See the main text for details

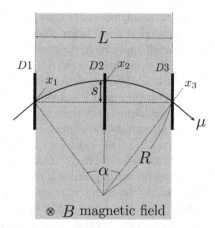

5.3.1   Momentum Measurement and its Resolution

Suppose a charged particle travels in the magnetic field of B (in Tesla) with the radius R (in meter). Also suppose we measure the charged particle positions by position-sensitive detectors $D1$, $D2$, and $D3$, as shown in Fig. 5.2. The momentum of the charged particle p_T (GeV) can be written as $p_T = 0.3 \times B(\text{Tesla}) \times R(\text{meter})$. Because the angle α in Fig. 5.2 is geometrically represented as $\alpha \approx \dfrac{L}{R}$, the depth of the arc called a sagitta (s in meter) of the particle trajectory can be expressed as

$$s = R \left(1 - \cos \frac{\alpha}{2}\right) \approx R \times \frac{\alpha^2}{8} = \frac{0.3BL^2}{8p_T}, \tag{5.1}$$

where L is the chord of the arc in meter. In case the track position at the three detectors is measured as $x_1 \pm \sigma_x$, $x_2 \pm \sigma_x$, and $x_3 \pm \sigma_x$ (with a common uncertainty of σ_x), the sagitta is $s = x_2 - \dfrac{x_1 + x_3}{2}$. The uncertainty of the sagitta is $\sqrt{\frac{3}{2}}\sigma_x$. Therefore, the momentum resolution can be represented as

$$\frac{\sigma_{p_T}}{p_T} = \frac{\sigma_s}{s} = \frac{\sqrt{\frac{3}{2}}\sigma_x}{s} = \frac{\sqrt{\frac{3}{2}}\sigma_x \cdot 8p_T}{0.3BL^2}. \tag{5.2}$$

In the same manner, in case s is measured at N points (N is more than about 10), the momentum resolution is represented as

$$\frac{\sigma_{p_T}}{p_T} = \frac{\sigma_x \cdot p_T}{0.3BL^2}\sqrt{\frac{720}{N+4}}. \tag{5.3}$$

From these calculations, you can see that the momentum resolution is proportional to the momentum of charged particle and the uncertainty of the position measurement ($\frac{\sigma_{p_T}}{p_T} \propto \sigma_x \cdot p_T$), and the inverse of the magnetic field and the square of the length of detectors. If you want to have better momentum resolutions, more detectors should be placed in a wider space where a stronger magnetic field is provided. This can be imagined if you draw the arc with 3 or more points in a limited space and estimate the curvature of its arc. For which can you estimate more precisely, an arc with a smaller radius or an arc with a larger radius?[1]

5.3.2 Momentum Scale Calibration

Going back to the calibration topics, a measurement of the trajectory of charged particles, or more specifically the sagitta, geometrical information of the tracking detector and knowledge of the magnetic field strength are necessary to derive the momentum, just as we have seen. Therefore, we don't really need the momentum scale calibration for the magnetic spectrometer in a sense, i.e. there are no conversions from a certain information to another such as the charge-to-energy conversion in a case of the energy measurement by a calorimeter.

But in most of the experiments, in situ calibration or correction of the momentum scale is performed for better accuracy and precision. A common technique is to make use of the known mass of some particles, for example, K_S, J/ψ or Z. The momentum scale of the reconstructed tracks is calibrated or corrected so that the peak position of the invariant mass distribution reconstructed from two tracks becomes the world average value[2] of K_S, J/ψ or Z. Figure 5.3 shows the invariant mass reconstructed

---

[1] The answer is "with a smaller radius" (under the same B and L).

[2] A physics quantity is measured by several different experiments. Such results are combined, that is, "averaged", for example, by the particle data group (PDG) [2]. Such combined results are called world average values.

Fig. 5.3 The invariant mass distribution reconstructed from two oppositely charged muons in the ATLAS experiment. Reprinted under the Creative Commons Attribution 4.0 International License from [3] © 2011 CERN for the benefit of the ATLAS Collaboration. The background contribution is subtracted. The data distribution is shown by either red or black dots, while the simulation by grey histogram

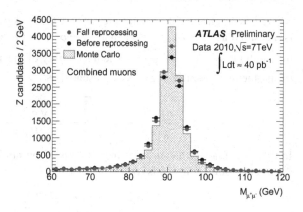

from two oppositely charged muons. As you can see, with this calibrated data, the peak position is consistent with the world average value of Z.

The particle used as the calibration target depends on the experiments because of the limitation of the available particles. The data sample with high purity is always preferable to avoid uncertainty due to the background. At the same time, the large data set is also preferable to reduce the statistical uncertainty. The experimentalist has to consider the optimal use of the various calibration samples.

This section was devoted to describing the momentum scale calibration or correction. But Fig. 5.3 shows the other important point which we would like to mention. It shows that the resolution depends on the alignment. As can be seen in Eqs. 5.2 and 5.3, momentum resolution has a linear dependence on the precision of position measurement for a track. Therefore, better alignment leads to better resolution. The figure shows that better alignment is used when the data was reprocessed.

5.4 Energy Calibration of Calorimeter

The energy calibration procedure for the calorimeter is classified into two steps. The first step is to calibrate each cell or channel, and the second is to calibrate the energy of the particle incident to the calorimeter, equivalent to the energy of the shower after clustering. These approaches are slightly different for the electromagnetic and hadronic calorimeters. Below, we discuss the concept of these two-step calibration procedures for the calorimeters.

5.4.1 Cell-by-Cell Calibration

In most cases, the energy information of the calorimeter is recorded as the digital number that is converted by an ADC from the detector output, typically the pulse height or charge created by the sensor. The goal of the cell-by-cell calibration is to find the relation between the energy deposit and the ADC count for each channel, which is a conversion factor. A set of the factors for all the cells are called calibration constants.

To get this calibration constant, the most powerful and a very common technique is to use a muon as the calibration source, because the muon in high energy physics experiment behaves as almost a minimum ionising particle (MIP) that deposits the constant energy per path length. The tracking system allows to measure the path length across the cell of the calorimeter, and hence to expect the energy deposit. In this way, one can obtain the ADC counts for unit energy. Only the muons can be this kind of calibration source, because the other charged particles evolve either electromagnetic or hadronic shower in materials, and their energy deposits are not constant. On the other hand, a muon deposits its energy just by ionisation loss, resulting in a rather constant energy deposit per unit path length. In the energy frontier collider experiments, muons decayed from Z bosons are one of the cleanest samples. They are isolated, i.e. there are no other particles nearby, and have high momentum. The higher the momentum. the multiple the scattering angles are smaller. This means that the error of estimating the path length is smaller. In addition, $J/\psi \rightarrow \mu^+\mu^-$ events are also used as the lower momentum calibration source.

The additional advantage of the usage of muons as the calibration source is the fact that high momentum muons, which are regarded as MIPs, are available in cosmic rays. We can have this ideal calibration source for free everywhere in the world, except the underground experimental facilities such as Super-Kamiokande, where the rate of muons is very small compared to the collider data.

In some cases, however, the in situ muon calibration may not be possible. In that case, the calibration results before installing the detector or assembling it into a big piece are used. For example, beam tests are employed, where the beam energy is precisely known. Or radioactive sources are also used because the energy spectrum of the emitted particles is well known.

In addition to the calibration with particles, a common approach is to prepare and use the artificially generated calibration source. For the detector whose output is lights, such as for scintillators, light flushers like lasers can be used to emulate the signal. For the detector whose output is electrical signals, such as liquid argon calorimeters, electrical test pulses to the readout electronics are often used. By using this kind of calibration sources, the relation between the detector output and the ADC count can be identified, although the relation between the detector output and the energy deposit is not. Still it is useful, much better than nothing, because, for example, relative gains within a detector can be monitored. This is of particular importance for the large-scale detectors where it is not an easy task to adjust the detector response for each individual channel. For this reason, most of the detector systems nowadays are equipped with such a calibration device that also works as the monitoring system of the detector performance.

5.4.2 Energy Cluster Calibration of Electromagnetic Shower

In principle, once the calibration constant for each cell or channel is obtained, one should be able to know the energy of the incident photon or electron to the calorimeter just by summing the energy of each channel associated with the energy cluster generated by the photon or electron. In practice, however, simple summing is not good enough for many reasons. For example, the energy deposited by the electromagnetic shower is much larger than that of muons, leading to the difficulty in the extrapolation to higher energy. Or there are dead materials among the active sensors consisting of a calorimeter, where missing energy due to the dead materials needs to be corrected. A different clustering algorithm may lead to a different energy sum even for the same event. Therefore, it is necessary to calibrate the detector in situ with either the electron or photon whose energy may be known without the calorimeter information. In this regard, the electron is a more user-friendly calibration source because other detectors rather than the calorimeter under calibration can detect the electron and measure its momentum. On the other hand, only the electromagnetic calorimeters can detect and measure the energy of photons precisely. This manifests the difficulty of in situ photon calibrations in collider experiments.

The most common and powerful technique using electrons exploits the fact that the electron's energy deposit ($\equiv E$) at a calorimeter should be equal to its momentum ($\equiv p$) at a tracker because an electron deposits all the kinematic energy at the calorimeter, and we can safely ignore the electron mass in the momentum region of our interests. Besides, for most of the momentum range in our interests, magnetic spectrometers consisting of charged particle tracking devices and magnets have better momentum resolution than that of calorimeters. Combining the above two facts, the momentum measured by the magnetic spectrometer can be a good reference for the electromagnetic scale. Commonly used is the E/p distribution where electrons or positrons make a peak at unity if the detector is properly calibrated.

Another calibration method, which does not rely on other detectors such as the magnetic spectrometers, makes use of the decay of particles whose masses are precisely known. The decays of $Z \rightarrow e^+e^-$ and $J/\psi \rightarrow e^+e^-$ are commonly used, where the calorimeter's response to the positron or electron is calibrated so that the reconstructed invariant mass gets closer to the world average value. The width of the invariant mass distribution should be narrower after the successful calibration. In the calibration using particle decays, we should be aware that the energy of particles in the calibration source is preferred to be close to the interesting range of your physics analysis to avoid a large extrapolation; in the above examples, the typical electron energy is of the order of 10 GeV in $Z \rightarrow e^+e^-$, while only a few GeV or less in $J/\psi \rightarrow e^+e^-$. It means that the former should be used for relatively high p_T electrons and the latter for low p_T. Finally, a decay chain with a high signal-to-noise ratio needs to be selected to avoid a possible bias due to the background.

The calibration method using mass, for example $\pi^0 \rightarrow \gamma\gamma$, can also be used for the photon calibration in principle. However, it is difficult to find a good decay chain which has enough statistics and covers the wide range of photon momenta. A lack of good calibration sources for photons is a common issue in many experiments. The widely used approach is to rely on the electron calibration because the detector response by electron and photon is similar at the first order. They both evolve the electromagnetic shower where the only difference is the initial depth of starting the shower. For precision, Monte Carlo simulation is often used to correct small differences in the detector responses between electrons and photons. Further, when experiments become more mature or have more statistics, rare processes can be the calibration source. An example is $Z \to \mu^+\mu^-\gamma$ where the photon is radiated off. The statistics is much smaller than that of $Z \to e^+e^-$. The momentum range of this photon is limited because the photon is radiated from a lepton from the Z boson decay. Still, $Z \to \mu^+\mu^-\gamma$ is used as calibration; to be precise, it is a validation tool to check the calorimeter response to photons.

5.4.3 Energy Cluster Calibration of Hadronic Shower

The energy-scale calibration for hadronic showers is much more complicated than that for electromagnetic showers for the following reasons.

First, the detector response to hadrons is different from that to electrons or photons, because of the different shower evolutions. Usually, all the kinematic energy of particles incident to a calorimeter is deposited in case of electromagnetic showers, meaning all the energy can be seen by the detector. On the other hand, some parts of the incident energy of hadrons are often lost because the nuclear interaction length (see Sect. 3.3.2) is relatively longer compared to the size or depth of the detector. Therefore, even if an electron and hadron have the same energy and hit into the same calorimeter, the "visible" energy may be different. Sometimes, this visible energy ratio of the electron and hadron with the same energy is referred to as the "e/h" ratio. With a few exceptions, most of the hadron calorimeters have $e/h > 1$, demanding a special correction in the calibration process.

Second, the fluctuation of energy deposits by hadrons is very large, while it is almost zero for the electromagnetic showers if the depth of the calorimeter is thick enough. The fluctuation comes from the fact that the hadronic shower sometimes creates π^0 that immediately decays into $\gamma\gamma$ and loses its energy by the evolution of electromagnetic showers. Hence in the case of having π^0 in the hadronic shower, the visible energy gets larger, and vice versa. Another reason for the large fluctuation is due to neutron production in the development of hadronic shower. In the case of charged hadron production such as proton, its kinematic energy can be detected as the energy deposit by ionisation, while low-energy neutron ($\ll 1$ GeV) does not have such an energy loss mechanism and is rather transparent in a calorimeter. Therefore, the visible energy is influenced by the number of produced neutrons. These are the main reasons why there is a rather large fluctuation in the energy loss of the hadronic shower. In addition, hadrons such as pion or kaon decay (semi-)leptonically, yielding neutrinos that are not detected by the calorimeter. The existence of neutrinos in the hadronic shower changes the total energy deposit in the calorimeter.

For the above reasons, in order to correctly deduce the energy of hadrons incident to the calorimeter, special care needs to be taken after the cell-by-cell calibration. In the collider experiments, it is rare to have a single hadron incident to the calorimeter. Instead, a jet (see Sect. 6.4) is the object handled by the calorimeter, which is an object defined by a human being. This means that the energy of a jet depends on the definition or actually on the clustering algorithm. For this reason, the treatment of jet energy calibration is described after introducing the jet reconstruction (see Sect. 6.4.4).

References

1. https://twiki.cern.ch/twiki/bin/view/AtlasPublic/InnerDetPublicResults2009
2. Zyla, P.A., et al.: [Particle Data Group]. PTEP **2020**(8), 083C01 (2020). https://doi.org/10.1093/ptep/ptaa104
3. ATLAS Collaboration, ATLAS-CONF-2011-046

Particle Identification

6

In this chapter, we discuss the method of particle identification, which is also called object identification because what we reconstruct or identify is usually not a particle but an *object* such as a charged particle trajectory, a jet that is a cluster of many particles, missing transverse energy and so on. We go through common objects that are widely used in high energy physics.

6.1   Tracking and Vertexing

A tracking refers to the reconstruction of a trajectory of a charged particle, or a track. Once we find such a trajectory under a magnetic field, we can measure the momentum of this particle through the curvature of the trajectory. The energy of a particle can be estimated with an assumption of the particle type, or with measurements related to the particle identification. Thus, the tracking allows us to obtain a 4-momentum vector, which is ultimately needed in data analysis. For this reason, the tracking capability is one of the most important features equipped in most of the detectors for the high energy physics.

The tracking can be divided into three parts: the measurement of the space hit points of the particles in the detector, the pattern recognition to the hit points to make a candidate track (referred to as track finding) and the fitting for the candidate track to get a smooth track, which is our best guess for the true particle trajectory. We discuss these three steps in the following subsections.

The collection of tracks in an event further allows us to find or guess, for example, the particle-particle collision points of the event, or decay positions of short-lived particles, such as K_S, Λ, b-hadrons, τ, etc. In either case, more than one particle appears from a common location, which we call a vertex. A vertexing refers to the reconstruction of the vertex using the collection of the tracks.

© The Author(s) 2022
K. Hanagaki et al., *Experimental Techniques in Modern High-Energy Physics*,
Lecture Notes in Physics 1001, https://doi.org/10.1007/978-4-431-56931-2_6

6.1.1 Space Hit Point

The tracking starts with searching for space hit points of charged particles in the detector. Charged particle deposits its kinematic energy in material by ionisation when passing through the materials. This results in some electrical signal or scintillation light, which will be converted to electrical signal in the end, in the sensor material. The sensor for tracking is usually segmented, allowing to know the hit position of the charged particle.

If the tracking device is pixelated, and had 100% efficiency without any fake hits, i.e. virtually a perfect detector, and only one charged particle existed in an event, the space hit point could be uniquely determined without any ambiguity, and there is no further discussion to find the space hit point. However, the real world is not so kind to us. The particle could penetrate through more than single segments due to the incident angle, resulting in multi-hits in a sensor. Moreover, many particles are often generated by particle-particle collisions or a beam hitting a target, and these particles may be overlapped and create multi-hits in a single sensor. Or there could be false measurements due to the inefficiency or the noise of the detector. First, in order to determine or estimate the space hit points, therefore, we have to apply clustering techniques to a set of raw hits in the sensor. Once we have a cluster of hits, then, the next step is to estimate the hit position of a particle from the cluster. Below we discuss the clustering techniques first, and then how to determine the hit position from the cluster.

Many tracking devices used so far are not pixelated. Instead, in many cases, they provide one-dimensional information from the wires in a chamber or strip electrodes in a silicon sensor. Therefore, we need to convert and obtain the space hit points from such one-dimensional measurements, which we will discuss at the last part of the following subsections.

6.1.1.1 Hit Clustering

Below we consider a position sensitive device in one dimension such as a silicon strip sensor, a wire chamber or a fibre tracker. A basic approach to the clustering is to group the consecutive hits. This is rather straightforward if the detector is perfect. All you need is just to set a (very low) threshold for each channel to define a single hit. In the actual experiments, however, there could be some dead channels or noisy channels where some particular channels have always a hit regardless the existence of a particle. Special treatments are needed for these kinds of deficits.

For example, in order to avoid a fake track caused by noise hits, we mask noisy channels, i.e. ignore such channels when making a cluster. Or if two silicon strip sensors can be a pair because they locate so closely, requiring two hits on the pair can reduce the fake hits significantly. In contrast, a known dead channel is sometimes treated as if it would have a hit when the adjacent two channels have hits.

In addition to the deficit such as noisy or dead channels, another type of care is needed when a large size cluster is generated. A large cluster is possible when a particle hits a sensor obliquely. In this case, energy deposit and hence signal size in each channel is small. So the threshold to define a hit needs to be set low to reconstruct this type of tracks. But of course such lower thresholds can cause higher noise rates. Therefore, an optimisation of the thresholds is always a key to have higher efficiency with lower fake hits.

6.1.1.2 Hit Point Determination

The method to estimate the hit position depends on how the signal information in each channel is stored. In the so-called "binary readout", where only the location of channels with the signal above a threshold is recorded, all we can do is just to take the average of the positions of hit channels. For example, if only one hit is found in a silicon strip sensor, the location of the strip with the hit is regarded as the incident position of a particle. If there are two hits, the particle incident position is regarded as the middle of hit strips, and so on. The position resolution of such binary readout scheme becomes $\frac{d}{\sqrt{12}}$, where d is the size of the segment of each channel, as discussed in Sect. 4.2.5.

In case the size of the collected charge at each channel is recorded by some means, the centre of gravity in terms of the collected charge can be considered to be the particle hit position. Even more sophisticated approach, such as making use of a lookup table, is sometimes used. In either way, the position resolution can be improved compared to the binary readout scheme. Let's assume that we now measure position along the x-axis with the silicon strip detector whose strip pitch is d, where we have two hits. Also, assume Q_L and Q_R be the collected charge at the two strips. With the centre of gravity method, x, which is the particle incident position defined as the distance from the right side strip, can be measured to be

$$x = d \times \frac{Q_L}{Q_L + Q_R} \equiv d \times \frac{Q_L}{S} \ ,$$

where S is the total charge accumulated by the two strips. Assuming the accumulated signal charge, S, is much larger than the noise, the uncertainty of measuring x can be written as

$$\delta x = d \times \frac{\delta Q_L}{S} \sim d \times \frac{N}{S} \ ,$$

where N is the noise. Since the binary readout gives us the position resolution of $d/\sqrt{12}$, the device with the signal-to-noise ratio (S/N) greater than $\sqrt{12} \sim 3.5$ has the advantage of using the analog information of the charge (Q). In many detectors including the silicon strip sensors, it is very common to have the S/N greater than 10 or 20. Hence, adding the analog information usually improves the position resolution.

6.1.1.3 From One-Dimensional Measurement to Space Point

In many cases, the tracking detector consists of devices that are capable of sensing only one-dimensional hit information, although it becomes popular to use the device with pixelated sensors recently. For example, suppose that there are parallel wires, which make a plane, inside the gas volume. By detecting a signal from the wire,

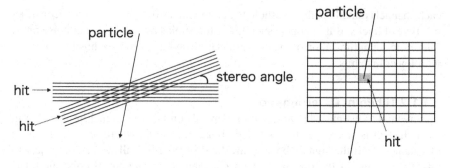

Fig. 6.1 Left: One-dimensional measurement by two layers with stereo angle gives two-dimensional measurement. Right: Two-dimensional measurement by pixelated device

one can identify the position of the particle hitting on the plane in the direction perpendicular to the wires, but not in the direction along the wires. Adding another layer in parallel to the original plane does not change the situation. Instead, placing another layer with some tilt angle with respect to the wires in the other layer allows us to get two-dimensional hit information. This tilt angle is referred to as stereo angle (Fig. 6.1: left).

The set of two layers with the stereo angle can provide two-dimensional measurements of a hit position on a certain plane. Since the three-dimensional space hit point is needed to reconstruct particle trajectory, one must define the plane of the two-dimensional measurements. In case the pixel type sensor is used, the plane is defined to be the sensor. In the right of Fig. 6.1, for example, x and y positions are determined by the location of pixel with hit, while the z is the plane of the sensor. Of course the sensor has finite thickness, and hence the "plane" must be pre-defined arbitrarily, such as the front surface or middle of the front and back side surface, etc. In case the configuration of two stereo sensor layers, such as two planes of wires (left of Fig. 6.1), the middle of two layers is often defined to be the space hit point plane.

The simple or natural idea to get two-dimensional hit information from two one-dimensional measurements is to have the stereo angle to be 90°. In many of the applications, however, it is difficult to have large stereo angle because of the geometrical constraints of the detector, especially in the collider detectors where the inactive materials such as cables must be minimised for 4π acceptance coverage. (In contrast, the fixed target experiment may not face such space constraints and can easily achieve 90° stereo angle.) To minimise the inactive materials, it is preferred to have the readout electronics including cables at only one end of the detector. Therefore, many tracking detectors that are actually used in the collider experiments have very small stereo angle so that the signals from two planes are read out through the same direction with a cost of losing the position resolution along the wire direction.

6.1.2 Track Finding

Once a set of space hit points is identified, the next step is to form the candidate tracks from the hit points. The idea itself here is simple, i.e. select the space hit points based on the track hypothesis where one needs a modelling of the charged particle trajectory. If only one particle travels inside the tracking volume without any noise hits, there would be no further discussions. All you need is to just connect the hit points. However, usually, there are many particles creating many hit points, resulting in very complicated hit pattern, plus noise hits. Given a particle trajectory model, the human eyes are very good in the pattern recognition, i.e. you can pick up a set of hit points to form a track. In the old days when using the emulsion or the bubble chambers, it was actually the eyes that worked as the track finder. But in most of the high energy physics experiments these days, one must use computers to find tracks through tons of the hit points, because of the very high rate of data acquisition. This means that algorithms for track finding have to be provided.

The first thing to consider is the modelling of charged particle trajectories. In the collider experiments, since the magnetic field exists in the tracking volume to measure momentum in most of the cases, the trajectory would be basically a helix. On the other hand, it is a straight line and possibly plus a curvature on the single plane, which is perpendicular to the magnetic field for measuring momentum in many of the fixed target experiments. There may be more models depending on each experiment. Anyhow, the point here is that the pattern recognition runs with a hypothesis on how a charged particle travels in the tracking volume.

The core of track finding is to select a set of hit points based on a given track modelling. There are mainly two groups of the track finding algorithms, in general, local and global methods, which we will describe below.

6.1.2.1 Local Method

The concept "local" means that the algorithm tries to find single track first, and then search for the next one once the first one is found. Until no possible candidates are found anymore, this procedure is repeated. In this way, multi-tracks are found sequentially.

The local method starts from finding or clustering an initial seed from the hit list. Suppose the tracking in the collider experiments such as the ATLAS experiment. The tracking algorithm picks up a hit on either the innermost or outermost layer. Here let's suppose that the algorithm looks for a track from the inside to the outside, i.e. the hit on the innermost layer is randomly picked up first. Then the hit on the next outer layer is searched, based on the hypothesis that the track comes from the proton-proton interaction point with a helix trajectory, where the rough estimate of the interaction point needs to be provided. The segment formed by the hits on the innermost and the adjacent outer layer is examined if it matches with the track hypothesis with some criteria. The survived segment becomes the track seed. Once the seed is found, the track candidate is extrapolated to the next outer layer and examined again if it has a hit. This procedure will be repeated until the track candidate reaches the outermost

Fig. 6.2 An example of the track and space hit point distribution in the plane perpendicular to the beam axis in the collider experiment without the magnetic fields. The coordinate origin is the collision point

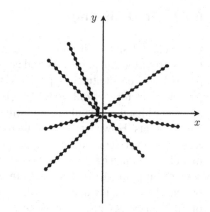

layer of the tracking volume. The examination is often based on the χ^2 testing or Kalman filter.[1]

Whether starting from inner to outer or from outer to inner depends on the algorithm. Because the inner layer has denser hits than the outer, starting from the outer allows to avoid unnecessary trials and hence save the CPU time. It also can reconstruct a track from the secondary vertex whose position is in the tracking volume. This type of secondary vertex may appear from the long lived particles such as K_S. On the other hand, the algorithm starting from the inside has the advantage in the efficiency by trying for any possibilities, even though it costs CPU time. In ATLAS, both algorithms are used in parallel. The resultant output, the track candidates from both the algorithms are fed into the next stage of the tracking, the track fitting.

After finding a track candidate, the space hits points that are used to reconstruct the track candidate are removed from the hit list for the next track candidates in many algorithms. Then the algorithm starts to search for the next track candidate using the updated hit list and continues the same procedure until no more candidates, which satisfy the selection criteria are found. In the dense experimental conditions such as the hadron colliders; however, it may be also a possible option to leave some hits on the list, even though they are already used, for the redundancy. The choice is up to you.

6.1.2.2 Global Method

The concept "global" means that the algorithm tries to find all tracks at a time from the list of the space hit points in contrast to the local method. There are other types of algorithms that are completely different from the one so far discussed. Actually, the variety of those algorithms is wide. So we just pick up and introduce two widely used algorithms.

---

[1] The Kalman filter is an algorithm to estimate the unknown status based on a series of observed measurement and its error. For example, see Ref. [1] for the detail.

Histogramming method For the illustration purpose, suppose we think of finding tracks in the collider detector without any magnetic field. Suppose also the particle-particle collision point (collision point in short) is known. Figure 6.2 shows the example of the space hit points projected onto the plane perpendicular to the beam direction. If you plot the azimuthal angle, ϕ, of the hit points on Fig. 6.2, the ϕ distribution will have the seven peaks, each corresponding to each track. By making such a histogram, one can identify the group of hit clusters in an event at a time. This is the basic idea of the histogramming method. In the actual application with the magnetic field, the coordinate transformation is carried out. Suppose x and y are the values in the original coordinate, then

$$u = \frac{x}{x^2 + y^2}$$

$$v = \frac{-y}{x^2 + y^2}$$

will produce straight lines in the (u, v) plane, ending up with the simple histogramming method which handles the straight lines.

Hough transformation Another famous example of the global method is the Hough Transformation, which is widely used in the digital imaging process. Suppose we have a line on $x - y$ plane. Given r is the distance between the line and the coordinate origin, and θ the angle between the x-axis and the normal of the line in problem, the line satisfies the following equation:

$$r = x \cos \theta + y \sin \theta$$

This means that any points (x_0, y_0) on the line satisfy

$$r(\theta) = x_0 \cos \theta + y_0 \sin \theta \ .$$

Therefore, the lines passing the point of (x_0, y_0) can be represented by a sine curve on the (r, θ) plane which we call a Hough space. Or the transformation of a set of lines on single point on the original $x - y$ plane to the sine curve on the Hough space is called Hough transformation. Let's now assume that we have five points on the straight line on the $x - y$ plane. A set of the straight lines passing each point is transformed to the corresponding sine curves on the Hough space. Since there are five points now, there are also five sine curves after the Hough transformation. Because the five points on the $x - y$ plane are on a common straight line, which we would like to find, there should be a crossing point of the five sine curves on the (r, θ) plane. The crossing point (r_0, θ_0) gives us the straight line on $x - y$ plane, i.e. $r_0 = x \cos \theta_0 + y \sin \theta_0$ is the expression of the line where the five points are on. This is how the Hough transformation allows to determine a line from a set of the space hit points.

The transformation shown above is for the straight line. It is possible to select a different transformation suitable for your application, for example, the curves such as the charged particle trajectories in magnetic field. In this case, assuming the trajectory to be a helix or a circle in the plane perpendicular to the magnetic field, the parameters representing the circle are the coordinate of the centre, x and y, and the radius, hence there are three parameters instead of two in the case of the straight line, r and θ. Then by performing the similar Hough transformation to the three-dimensional space, one can obtain a curved surface for each single point on $x - y$ plane. By repeating the Hough transformation from all the measured points, we get a set of the curved surfaces. The crossing of the curved surfaces represents the circle, which is the one we want to determine on the $x - y$ plane, or actually a track.

6.1.3 Track Fitting

The final step of the tracking is the fitting of the space hit points that are associated with each track candidate to the track modelling. The concept of the track fitting is rather simple, i.e. it is the least squares method to minimise the difference between the measurements and the track hypothesis. More specifically, χ^2 is defined to be

$$\chi^2(\theta) = \sum_{i}^{N} \frac{(y_i - f(x_i; \theta))^2}{\sigma_i^2} \quad ,$$

where y_i is the set of measurements at x_i, $f(x_i; \theta)$ the prediction of the track based on some trajectory modelling, σ the measurement error, and θ the parameter which you want to obtain by minimising the χ^2. In case of the collider experiments, a trajectory of charged particles is modelled by helix that consists of five parameters.

Assuming each measurement is described by Gaussian p.d.f., which is the case in many measurements, the χ^2 can be written as

$$\chi^2(\theta) = \sum_{i,j=1}^{N} ((y_i - f(x_i;\theta))(V^{-1})_{ij}((y_j - f(x_j;\theta)) \ ,$$

where V is the covariance matrix. This can be further written in general matrix notation as

$$\chi^2 = (\mathbf{y} - \mathbf{f})^T V^{-1}(\mathbf{y} - \mathbf{f}) \ ,$$

where $\mathbf{y}$ is the vector of the measurements, and $\mathbf{f}$ is the predicted value. The track fitting searches for the parameters θ to minimise the χ^2. There are varieties in the approaches for this minimisation. The details of the mathematical treatments can be found elsewhere (see Ref. [2] for example). Instead we discuss a few points that need to be considered particularly in high energy physics.

- The Kalman filter is widely used recently because it can naturally handle some effects due to the interaction of a particle and the material in the tracking detector, such as Coulomb multiple scattering and/or energy loss in the tracking device.
- The speed is crucial. This is especially true for the experiments with high rate and high multiplicity environment such as the hadron colliders. Not only the efficiency and/or resolution but also the speed needs to be optimised in the algorithm.

After the success of converging the track fitting, one can finally obtain the reconstructed tracks. As all the track parameters are determined at this point, one can deduce the momentum of the reconstructed tracks or the particles at arbitrary position. At the collider experiments, the momentum at the collision point is the interesting quantity we want to know in most cases. In addition, the complete track parameters allow to predict or extrapolate the particle trajectories outside the tracking volume, which is sometimes important in the particle identification such as the electron or muon identification, as shown in later in this chapter.

6.1.4 Vertex Finding

It is important to know the location of the collision point because it is where the particle reaction in interest happens, and hence the momentum vectors of the generated particles are defined. Therefore, the measurement of the collision point event-by-event is a crucial ingredient of the physics analysis. The analysis handling neutral particle is of particular importance because the momentum vector of the neutral particle cannot be defined without knowing the location of the collision point.

In the collider experiment, the collision point is often called as a primary vertex. There is also another type of vertex. For example, b-hadrons generated by the collisions can travel a few mm at LHC, and decay subsequently, creating a vertex at the decay point of the b-hadron because the decay products emerge and form a kink.

This type of a vertex, caused by a decay of some particles, is called as a secondary vertex.

The concept of the vertex finding is simple, i.e. after the tracking one extrapolates more than one reconstructed tracks to the direction where the vertex is expected to exist. The intersection of the extrapolated tracks can be a vertex.

The actual vertex finding starts to pick up all reconstructed tracks with some selection criteria, which assures the quality of the tracks. The tracks are then examined which one should be associated with the same vertex candidate by the least squares fitting, or equivalently, Kalman filtering. The track that significantly worsen the χ^2 value is removed from the association of the vertex or down-weighted in the fitting. The latter technique is called as the adaptive vertex fitting. Until stabilising the χ^2 value, the testing of the association of the tracks continues. The fitting after this removal process gives us the best possible vertex estimate.

In order to improve the track reconstruction and the vertex finding, the track fitting is repeated after finding the vertex, and then vertex fitting again. This recursive process is of particular importance in the dense environment such as the hadron colliders where not only tons of tracks exist but also many interaction/collision points, hence many vertices, exist per bunch crossing.

At the hadron collider experiments, there are many interactions occurred per bunch crossing as just mentioned. At the LHC, for example, more than 20 interactions occur in average as shown in Fig. 6.3, where 65 vertices are found, while the bunch length

Fig. 6.3 Event display obtained from the real data in the ATLAS experiment. Reprinted under the Terms of Use from [3] ATLAS Experiment © 2022 CERN. All rights reserved. The reconstructed tracks are shown in the lines, and the vertex by the circles. There are 65 proton-proton interaction points on top of the one where Z boson is produced and decayed into dimuon. This dimuon is shown in yellow lines

is 7.5 cm. Out of these 65 interaction points, we have to find out the vertex where our interesting event is generated as the next step. In the example of Fig. 6.3, two muons appeared from one of the 65 vertices are identified to be consistent with Z decay. Since such interesting events often have high momentum transfer of the colliding partons, called hard scattering (see Sect. 2.5), the selection of the vertex is based on some measures that represent the momentum transfer of the collision. For this reason, the sum of p_T of the tracks, or the number of tracks associated to each vertex, is frequently used. Note that the hadron collider people sometimes call only this hard scattered interaction point as the primary vertex, although the primary means vertices generated by the collisions and the secondary by the particle decays in the original definition. They call the other vertices induced by the collisions as the primary vertex "candidates" or pile-up vertices. Readers should be aware of this difference.

Once we know the location of the primary vertex, or the hard scattered vertex, a momentum vector of the charged particle reconstructed by tracking is recalculated to obtain the one with respect to the primary vertex. This is one of the information ultimately necessary in physics analysis using charged particles.

A special care needs to be taken to find secondary vertices, as the location of the secondary vertex is not known a priori, in contrast to the primary vertex finding where the collision point is known to some degree. The basic procedure is the same as the primary vertex search. Only the explicit difference is that the tracks not associated with the primary vertex are the candidates to form the secondary vertex. The idea here is simple, but this selection is sensitive to the performance of the secondary vertex finding, and also on the primary vertex reconstruction. For example, if the selection criteria are tight, i.e. if you select only the tracks whose impact parameter is far enough away from the primary vertex, the reconstruction efficiency of the secondary vertex would be poor, while the reconstructed primary vertex position is free from the possible bias due to the tracks emerged from the secondary vertex. Therefore, the optimisation based on the physics requirements is of particular importance. In the end, the secondary vertex finding allows us to identify the location of decay position of K_S, Λ, b-hadrons, and so on, which is the important ingredient of physics analysis.

6.2 Electron and Photon

6.2.1 Interactions with Materials

Electrons and photons lose their energy by developing a characteristic shower in the electromagnetic (EM) calorimeter, which is called an EM shower due to the cascade process of the bremsstrahlung ($e + $ materials $\rightarrow e\gamma$) and the e^+e^- pair production ($\gamma + $ materials $\rightarrow e^+e^-$) as shown in Fig. 6.4. Such processes are based on

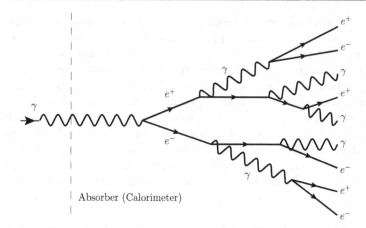

Fig. 6.4 Schematic views of a EM shower: a photon is injected into a calorimeter

the interaction of electrons and photons with materials of the absorbers, for example lead (Pb) in the ATLAS detector and crystal (PbWO$_4$) in the CMS detector.[2]

A charged particle like an electron interacts with electrons in atoms (electrons of molecules) of detector materials through the EM interaction. A charged particle ionises or excites atoms. This is why a charged particle loses energy in materials. This is called ionisation loss. This energy loss can be described by the Bethe-Bloch formula. In addition, a charged particle radiates photons when it is decelerated in materials, which is called bremsstrahlung. The ionisation loss is dominant in a low energy region (low $\beta\gamma$) while the bremsstrahlung in a high energy region (high $\beta\gamma$). The energy at which the ionisation loss and bremsstrahlung energy loss is equal is called a critical energy. It is about 7.3 MeV for electrons in Pb.

There are three kinds of interactions for a photon with materials: the photoelectric effect, Compton scattering and e^+e^- pair production. In the energy of $> 2m_{electron} = 1.022$ MeV, the e^+e^- pair production is dominant. Once an e^+e^- pair is created, the interaction of an electron with materials can be applicable. This is why the detector response by a photon is similar with that by an electron at the first order as mentioned in Sect. 5.4.2.

The development of the EM showers stops at the critical energy. One of the key parameters to describe calorimeter performance is the radiation length X_0, which represents a length with which the energy of an electron becomes a factor of $1/e$ by passing materials. A unit of X_0 is g/cm^2 or cm: X_0 for Pb is 0.56 cm, X_0 for PbWO$_4$ 0.89 cm. Typical EM calorimeters have at least $20X_0$ (ideally $25X_0$) to stop EM showers. Materials with smaller X_0 can stop EM showers with a small space.

The main materials for electrons and photons to lose their energy are those of calorimeter detectors. In most of actual collider experiments, however, there are

---

[2] The ATLAS EM calorimeter is a sampling calorimeter, where the absorber is Pb and the detector is based on liquid argon and the CMS EM calorimeter is a homogeneous calorimeter, where the absorber and detector parts are made of the crystal (PbWO$_4$).

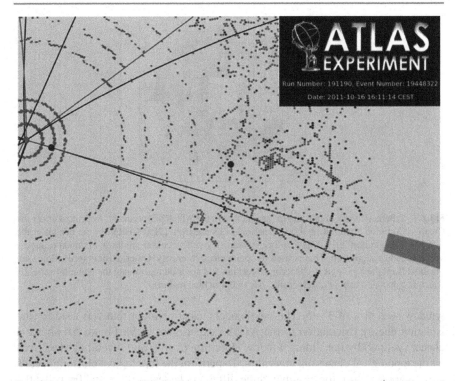

Fig. 6.5 A photon interacts with the inner tracking detector and is converted into e^+e^- (blue and red curves) in the ATLAS detector. The conversion vertex is shown with a brown point. Reprinted under the Creative Commons Attribution 4.0 International License from [4] © 2011 CERN for the benefit of the ATLAS Collaboration

materials in front of the calorimeter where the bremsstrahlung, e^+e^- pair production etc. are possible, for example, a beam pipe and inner tracking detectors. An electron can be scattered by the Coulomb force (called the multiple scattering) or radiate photons via the bremsstrahlung. A photon can be converted into an e^+e^- pair as shown in Fig. 6.5, which produces oppositely charged particles with a zero-opening angle and imbalance in momenta. In this case, a positron and an electron are detected in the calorimeter instead of a photon. Such a photon is called a *converted photon*. The position where such a conversion happens, which is called a *conversion vertex*, can be obtained from these two charged tracks reconstructed in the inner tracking detector. When photons themselves are detected in the calorimeter without the conversion, they are often called *unconverted photons* to distinguish from converted photons.

6.2.2 Reconstruction

Electrons and photons are reconstructed by clustering cells of the EM calorimeter, where their energies are deposited in each cell. Typical algorithms are: the sliding window algorithm [5], the topological-clustering algorithm [6] etc. The reconstruction of electrons and photons is based on the sliding window algorithm with a sliding

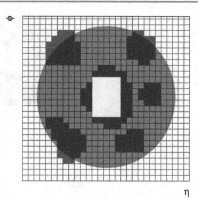

Fig. 6.6 Schematic view of clusters for an electron in the ATLAS experiment. Reprinted under the Creative Commons Attribution 4.0 International License from [7] © CERN for the benefit of the ATLAS collaboration 2019. The 5×7 cells shown by yellow colour are those obtained using the sliding window algorithm. This is used to obtain electron energy. Other clusters (red colour) are obtained from the topological-clustering algorithm and are used to evaluate the isolation variable, which is calculated using clusters inside $\Delta R = 0.4$ (a blue region)

window seed size of 3×5. The topological clustering algorithm was used for the isolation energy calculation in the ATLAS experiment [7,8]. Figure 6.6 shows a cluster (yellow) by the sliding window algorithm (5×7) and several clusters (red) by the topological clustering algorithm. To perform the topological clustering, cells are categorised into, for example, three different classes: 4σ, 2σ and 0σ cells: the 4σ cells are those having energy of four or more times larger than their expected noises (σ), the 2 $(0)\sigma$ cells are those having energy of $2 - 4$ $(< 2) \times \sigma$. Then, in the step of the clustering, one of 4σ cells is selected as a seed cell and its neighbouring 4 or 2σ cells in the three spatial directions are connected until there are no neighbouring 4 or 2σ cells. Then, all the surrounding 0σ cells are finally connected to have clusters as electron and photon candidates.

For electrons, they can be also reconstructed through charged tracks using the inner tracking detector information (Sect. 6.1) since they are charged particles. Clusters matched to a charged track are classified as electrons and those not matched to any charged tracks are as unconverted photons. The momentum of the matched charged tracks is recalculated taking into account possible energy loss due to the bremsstrahlung in the detectors in front of the calorimeter. A cluster matched to a track pair from a reconstructed conversion vertex or a single track that has no hit in the innermost layer of the inner tracking detector is classified as a converted photon. The energy calibration of electrons and photons is explained in Sect. 5.4.2.

6.2.3 Identification

Reconstructed electrons and photons are candidates of true electrons and photons, respectively. Other particles, which are background for electrons and photons, can be also reconstructed with the same algorithm mentioned above. Such particles are

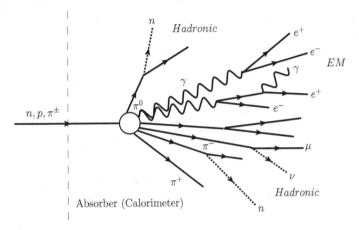

Fig. 6.7 Schematic views of a hadronic shower: a hadron (n, p, $\pi^{\pm}$, etc.) is injected into a calorimeter

dominated by jets, which are explained in detail later (Sect. 6.4). The origin of jets is a gluon and a quark, which produce a set of particles after hadronisation. A jet or a hadron can develop a hadronic shower in the calorimeter. The hadronic shower, which is shown in Fig. 6.7, has two components: hadronic and EM components. The hadronic component produces charged pions, charged kaons, protons, neutrons etc. through the hadronic interaction (strong interaction).[3] In addition, neutral pions are also produced but they are observed as photons since the lifetime of neutral pions (8.5×10^{-17} s) is very short and they immediately decay into two photons. This is the EM component of a hadronic shower.

Electrons and photons can be separated from jets and hadrons using the differences between an EM shower and a hadronic shower: lateral (=transverse) and longitudinal shower developments are different. For the lateral shower shape, an EM shower is relatively narrower than a hadronic shower since the constituents of a jet ($\pi^{\pm}$, $K^{\pm}$, p, n, γ etc.) are spread. This is the case even for a single hadron, where the hadronic shower can become wider with producing neutrons etc. For the longitudinal shower shape, a hadronic shower is developed into the hadronic calorimeter, in other words, the shower cannot stop in the EM calorimeter. For example, when there are some longitudinal layers in the EM calorimeter like the ATLAS detector, the energy deposited in outer layers of the EM calorimeter is larger for jets and hadrons than for electrons and photons. Variables for the identification can be defined using cells of calorimeters. Such variables are called shower shapes variables, for example, shower widths, ratios of energy deposited in different layers of the calorimeter, etc. Figure 6.8 shows four variables for electrons in the ATLAS experiment: $w_{\eta 2}$ and R_{η} represent a kind of narrowness in the lateral direction, and R_{had1} and f_3 for the shower development in the longitudinal direction. Since more than 10 variables of

---

[3] Muons and neutrinos are produced from the decay of charged pions via the weak interaction: $\pi^{\pm} \rightarrow \mu^{\pm}\nu$. The energy of these particles is largely undetected in the calorimeter.

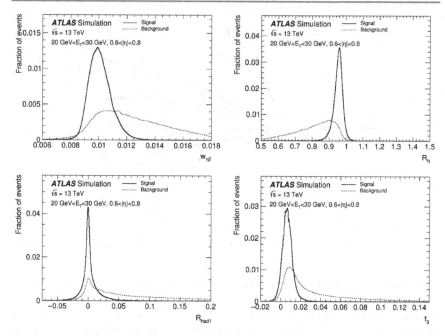

Fig. 6.8 Distributions of two shower shapes for the electron identification from the ATLAS MC simulation studies. Reprinted under the Creative Commons Attribution 4.0 International License from [7] © CERN for the benefit of the ATLAS collaboration 2019. $w_{\eta 2}$ and R_η are a shower width and a ratio of the energy in 3×3 cells over the energy in 3×7 cells in the second layer of the EM calorimeter, respectively. R_{had1} is a ratio of the transverse energy (E_T) in the first layer of the hadronic calorimeter to E_T of the EM cluster. f_3 is a ratio of the energy in the third layer to the total energy in the EM calorimeter. Signals are electrons from Z and J/ψ decay and backgrounds are from electron candidates from multijet production, γ+jets etc.

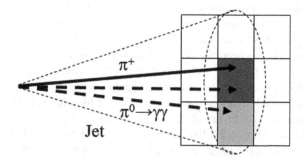

Fig. 6.9 Schematic view of a fake electron from a jet: a charged pion overlaps with photons from a neutral pion decay inside a jet

shower shapes and tracks (if necessary) are used, so-called a multivariate analysis technique such as a combined likelihood, neural network, or boosted decision tree is adopted. Possible reasons of misidentification for electrons and photons using the shower and track variables are given below.

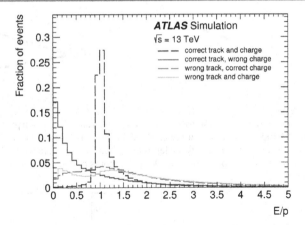

Fig. 6.10 E/p distributions for electrons (signal) and hadrons (background) from the ATLAS MC simulation studies. Reprinted under the Creative Commons Attribution 4.0 International License from [7] © CERN for the benefit of the ATLAS collaboration 2019. E is energy measured in the calorimeter and p is momentum measured in the inner tracking detector

Jets can be misidentified as electrons (called *fake electrons*), for example, because a charged pion overlaps with photon(s) from a neutral pion decay, a η decay and so on inside a jet. This is illustrated in Fig. 6.9. One of useful discriminating variables for this type of fake electrons is E/p as shown in Fig. 6.10, where E is energy measured in the calorimeter and p is momentum measured in the inner tracking detector. In case of true electrons, it should be close to 1 because E and p should originate from a same object but in case of jets (fake electrons) there is no clear correlation between E and p because different particles can contribute to E or p. This variable is included in the electron identification.

Not only jets but also other objects such as τ-jets and converted photons are misidentified as electrons. τ-jets are misidentified when it decays hadronically to one charged particle, so-called one-prong (*ex.* $\tau \rightarrow \pi^{\pm}\pi^{0}\nu$) with the same reason as jets, that is, the overlap between $\pi^{\pm}$ and γ from π^{0}. A simple τ-veto algorithm is applied: electron candidates that are highly identified as τ-jets in a τ identification are rejected. For the converted photons, a cluster has a possibility to have a matched charged track when one of the charged particles is not reconstructed. To reduce such misidentification (see Fig. 6.11), a track is required to associate to a primary vertex using impact parameters since tracks from a conversion vertex have large impact parameters. In addition, a hit in the innermost layer of the inner tracking detector is required for electron candidates.

Jets are misidentified as photons (called *fake photons*), for example, when a neutral pion from a jet carries most of energy of the jet. This is illustrated in Fig. 6.12. In principle, two photons should be observed inside a jet because a neutral pion decays into two photons. To separate a single photon from a set of two photons, the finely segmented first layer is used in the ATLAS experiment as shown in Fig. 6.13. A single cluster is observed for a photon (left) but two clusters for a π^{0}.

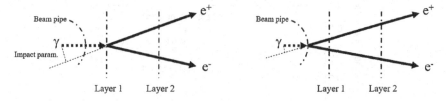

Fig. 6.11 Schematic views of conversions at the first layer (left) and at the beam pipe (right). When two tracks are reconstructed, both cases are categorised into conversions. On the other hand, in case one of the tracks is misreconstructed, only the left is taken as a conversion to reduce fake converted photons. No hit in the first layer of the inner tracking detector is required for conversions

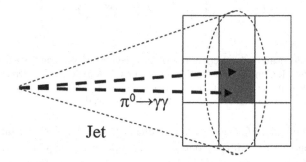

Fig. 6.12 Schematic view of a fake photon from a jet: most of jet's energy is carried by a neutral pion

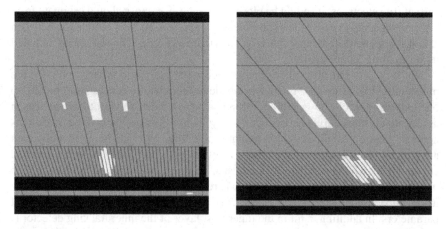

Fig. 6.13 Energy deposits in the three layers in the ATLAS EM calorimeter for a photon (left) and two photons from a π^0 (right). Reprinted under the Terms of Use from [9] ATLAS Experiment © 2022 CERN. All rights reserved. Photons are injected from the bottom to the top. The energy deposits are shown by yellow. In the right figure, two groups of the energy deposit are observed in the first layer (fine segmentation) of the calorimeter

6.3 Muon

Muons can be produced via decays of Higgs bosons, W/Z bosons, quarks and new particles such as SUSY. Therefore, the reconstruction and identification of the muons with good quality in the wide range of momentum and solid angle are key to many of the most important physics in the energy frontier experiment.

The muon belongs to the second-generation lepton. The characteristics are similar to the electron except for the mass, with an electronic charge of $-e$, a spin of 1/2, a mass (m_μ) of $105.6583715 \pm 0.0000035$ MeV. Muons hardly make either electromagnetic or hadronic shower in our energy regime, but decay into an electron, an electron anti-neutrino, and a muon neutrino via the weak interaction. Therefore, the mean lifetime of the muon (τ_μ), even flying in material, is very close to that in vacuum, 2.1969811 ± 0.0000022 μs, which is relatively long. Muons with the momentum (p_μ) of 1.0 GeV can pass through around 70 m at a period of the mean lifetime in the laboratory frame;

$$c\tau_\mu\beta\gamma = c\tau_\mu\frac{p_\mu}{m_\mu} \approx 3 \times 10^8 \text{ (m/s)} \times 2.2 \times 10^{-6} \text{ (s)} \times \frac{1.0 \text{ (GeV)}}{0.105 \text{ (GeV)}} \approx 70 \text{ (m)}$$

(6.1)

where the c is a velocity of light, $\beta = \dfrac{v}{c}$ is a ratio of the velocity of the muon to the light velocity, and $\gamma = \dfrac{1}{\sqrt{1-\beta^2}}$ is a Lorentz boost factor.

Considering these unique characteristics of the muon, in the collider experiments, muons can be detected by charged particle detectors located at both inside and outside of calorimeters. In this section, the muon identification and reconstruction in the collider experiments are described using the ATLAS detector as an example.

6.3.1 Muon Momentum Measurement

Muon reconstruction and identification in ATLAS relies on inner tracking detector, described in Sect. 3.3.2, and muon spectrometers (MS). The track reconstruction is first independently performed in inner tracker and MS. The information from both of them is then combined to form the muon tracks that are used in the physics analysis.

6.3.1.1 Effect of Multiple Scattering

When a muon passes through the large volume of the materials in the detectors, the effect of the multiple scattering needs to be taken into account. The multiple scattering angle is regarded as the accumulation of the Rutherford scattering. The probability of single Rutherford scattering is in inverse proportion to $\sin^4\left(\frac{\theta}{2}\right)$, where the θ is the scattering angle. The scattering angle has a sharp peak at $\theta = 0$, meaning that θ is typically very small. The mean of the multiple scattering angle is statistically regarded as the accumulation of the small angle Rutherford scattering, shown as

$\langle \theta^2 \rangle = \sum_i \theta_i^2$. The multiple scattering angle approximately distributes in the Gauss distribution. The effect of the large angle scattering that also occurs for $\sin^4 \left(\frac{\theta}{2} \right)$ distribution is shown up in the tail of the distribution. The mean of the multiple scattering angle ($\theta_0 \equiv \sqrt{\langle \theta^2 \rangle}$) can be expressed as

$$\theta_0 = \frac{13.6 \, \text{MeV}}{\beta c p} \sqrt{\frac{x}{X_0}} \left[1 + 0.038 \ln \left(\frac{x}{X_0} \right) \right] \propto \frac{1}{p} \sqrt{\frac{x}{X_0}}, \tag{6.2}$$

where p and βc are the momentum and velocity of a muon, respectively, and x/X_0 is the thickness of the scattering medium in radiation length (X_0). If the uncertainty of the muon position measurements, σ_x in Eq. (5.2) or (5.3) dominated by the multiple scattering with the detector materials, the momentum resolution is independent of p_T as

$$\frac{\sigma_{p_T}}{p_T} \propto \sigma_x \cdot p_T \propto \theta_0 \cdot p_T \propto \frac{1}{p_T} \cdot p_T \approx \text{const.} \tag{6.3}$$

6.3.1.2 Contributions to Muon Momentum Resolution

The uncertainty of the position measurement σ_x usually comes from the accuracy of the hit position measurement limited by the detector characteristic, misalignment of detectors, multiple scattering in the detector, and fluctuations in the energy loss of the muons traversing through the material in front of the spectrometer. Figure 6.14 shows the contributions to the momentum resolution for the ATLAS MS as a function of transverse momentum [10]. The contribution of the multiple scattering is independent of the transverse momentum and dominated at moderate momentum ($30 < p_T < 300$ GeV), while the contributions of the hit position resolution (denoted as "Tube resolution and autocalibration" in Fig. 6.14) and the detector (chamber in this case) alignment are in inversely proportion to p_T and dominated at high momentum ($p_T > 300$ GeV). At low momentum ($p_T < 30$ GeV), energy loss fluctuations become dominant.

The ATLAS MS is designed to detect muons in the pseudorapidity region up to $|\eta| = 2.7$ and to provide momentum measurements with a relative resolution better than 3% over a wide p_T range and up to 10% at $p_T \approx 1$ TeV. In order to satisfy the requirements, the measurement precision in each hit by a muon track is required to be typically better than 100 μm, which can be roughly estimated by Eq. (5.2). The uncertainty of the alignment in the chamber positions is required to be at the level of 30 μm.

Figure 6.15 shows muon momentum resolution for MS alone and for the combined measurements by MS and inner tracker [10]. At low momentum ($p_T < 30$ GeV), the measurement by inner tracker is better due to better spatial resolution of silicon strip and pixel detectors. On the other hand, at high momentum ($p_T > 30$ GeV), the measurement by MS becomes better than inner tracker because the MS is stationed in a wider space, which means L is larger in Eq. (5.2).

Fig. 6.14 Contributions to the muon momentum resolution for the ATLAS MS as a function of transverse momentum. Reprinted under the Creative Commons Attribution 3.0 License from [10] © 1997-2022 CERN

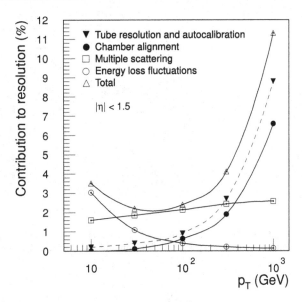

Fig. 6.15 The muon momentum resolution for the muon spectrometer alone and the combined measurements by the ATLAS MS and the ATLAS inner tracker as a function of the transverse momentum. Reprinted under the Creative Commons Attribution 3.0 License from [10] ATLAS Collaboration © 1997 CERN. The dashed curve is the resolution using only the inner tracker

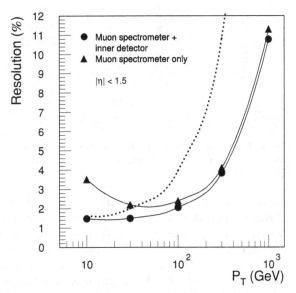

6.3.2 Examples of Muon Detectors

Because the muon spectrometers have to cover the wide surface area of the barrel and endcap of the cylindrical detector system, it is required to be robust, mechanically strong, and inexpensive as well as to provide the good momentum resolution and the high efficiency. Because muons give us clear signatures from physics of interests such as $H \rightarrow ZZ^* \rightarrow 4\mu$, the muon spectrometers are used as the trigger devices which provide fast information on momenta, positions and multiplicity of muons

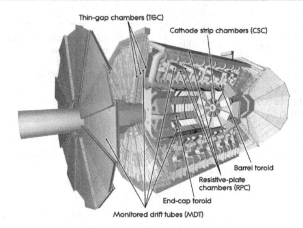

Fig. 6.16 The muon spectrometer for the ATLAS experiment. Reproduced by permission of IOP Publishing from [11] © IOP Publishing Ltd and SISSA. All rights reserved

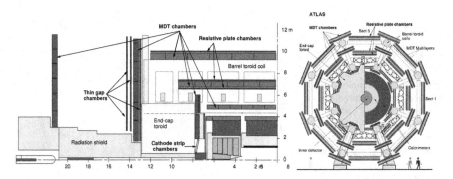

Fig. 6.17 The cross-section of the ATLAS muon spectrometer: r-z view (Left) and r-ϕ view (Right). Reprinted under the Creative Commons Attribution 3.0 License from [10] ATLAS Collaboration © 1997 CERN

traversing through the detector. This is called as the first level muon trigger, which makes a trigger decision within a few micro-seconds by a simple trigger logic on hardware. The gas detectors satisfy these requirements. For instance, the ATLAS MS, shown in Figs. 6.16 and 6.17, consists of the resistive plate chambers (RPC) and the thin gap chambers (TGC) to provide the fast muon trigger information and the monitored drift tube (MDT) chambers and the cathode strip chambers (CSC) to reconstruct muon trajectory precisely. The ATLAS MS divided into a barrel part ($|\eta| < 1.05$) and two endcaps ($1.05 < |\eta| < 2.7$).

Three large superconducting air-core toroid magnets provide magnetic fields with a bending integral of about 2.5 T·m in the barrel and up to 6 T·m in the endcaps in order to measure the muon momentum independently to the inner tracking system with the solenoid magnet (Fig. 6.18). In the following sections, as an example of the muon chamber, RPC, TGC, MDT used in ATLAS muon spectrometers, are introduced [11].

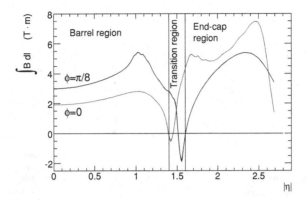

Fig. 6.18 The magnetic fields provided by ATLAS toroid magnet. Reproduced by permission of IOP Publishing from [11] © IOP Publishing Ltd and SISSA. All rights reserved

6.3.2.1 Resistive Plate Chamber

In the barrel region ($|\eta| \leq 1.05$), trigger signals are provided by a system of resistive plate chambers (RPCs). The RPC is a gaseous parallel electrode-plate detector providing a typical space-time resolution of 1 cm × 1 ns with digital readout. The mechanical structure of an RPC is shown in Fig. 6.19. Two resistive plates, made of phenolic-melaminic plastic laminate, are kept parallel to each other at a distance of 2 mm by insulating spaces. The gas gaps are filled with the gas of a mixture of $C_2H_2F_4$/Iso-C_4H_{10}/SF_6 (94.7/5/0.3). The electric field between the plates of about 4.9 kV/mm allows avalanches to form along the ionising tracks towards the anode. Since all primary electron clusters form avalanches simultaneously in the strong and uniform electric field, single signal is produced instantaneously after the passages of the particle. The intrinsic time jitter is less than 1.5 ns. The signal is read out via capacitive coupling to metallic strips, which are mounted on the outer faces of the resistive plates. The total jitter of RPC is less than 10 ns, which ensures to identify the proton bunch crossing of 25 ns and to produce fast trigger signals. The readout pitch of η and ϕ-strips is 23–35 mm. The η and ϕ strips provide the bending view of the trigger detector and the second-coordinate measurement, respectively. The second-coordinate measurement that cannot be done by MDT chambers (see Sect. 6.3.2.3) is also required for the offline pattern recognition.

RPC is made up of three stations, each with two detector layers. Two stations installed at a distance of 50 cm from each other are located near the centre of the magnetic field region and provide the low-p_T trigger ($p_T > 6$ GeV) while the third station, at the outer radius of the magnet, allows to detect the muon trajectory with larger curvature and to increase the p_T threshold to 20 GeV, thus providing the high-p_T trigger. The trigger logic requires three out of four layers in the middle stations for the low-p_T trigger and, in addition, one of the two outer layers for the high-p_T trigger (Fig. 6.20).

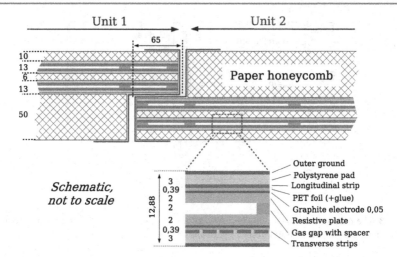

Fig. 6.19 Mechanical structure of an RPC chamber. Reproduced by permission of IOP Publishing from [11] © IOP Publishing Ltd and SISSA. All rights reserved. The unit of the number in the figure is mm

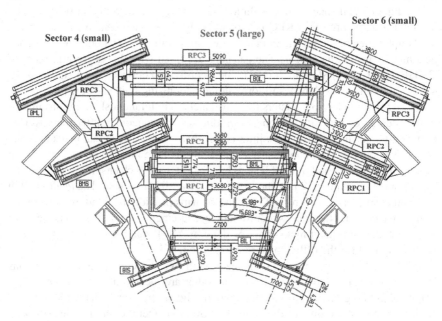

Fig. 6.20 Cross-section of the upper part of the barrel muon spectrometer. Reproduced by permission of IOP Publishing from [11] © IOP Publishing Ltd and SISSA. All rights reserved. Two stations of the RPC are below and above middle station of MDT chamber. Outer station is above the MDT in the large and below the MDT in the small sectors. Dimensions are in mm

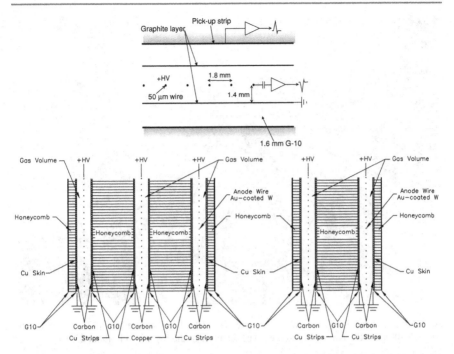

Fig. 6.21 TGC structure showing anode wire, graphite cathodes, G-10 layers and a pick-up strip, orthogonal to the wires (top) and cross-section of a TGC triplet and doublet module (bottom). Reproduced by permission of IOP Publishing from [11] © IOP Publishing Ltd and SISSA.

6.3.2.2 Thin Gap Chamber

In the endcap region ($1.05 \leq |\eta| \leq 2.4$), trigger signals are provided by a system of thin gap chambers (TGCs). TGC is multi-wire proportional chambers with the characteristic that the wire-to-cathode distance of 1.4 mm is smaller than the wire-to-wire distance of 1.8 mm, as shown in Fig. 6.21. The gas used is mixture of CO_2 and n-C_5H_{12} (n-pentane) (55 : 45). TGC is operational in quasi-saturated mode with a gas gain of about 3×10^5. The high electric field of the wires (around 2800 V) and small wire-to-wire distance allows us to measure the muon trajectory with a good time resolution and to identify the proton bunch crossing of 25 ns. The number of wires in a wire group varies from 6 to 31 as a function of η, in order to match the granularity to the required momentum resolution. The wire groups measure the η direction of the muon trajectory. Two of copper layers in triplet and doublet modules, which is marked as "Cu stripes" in Fig. 6.21, are segmented into readout strips to read the azimuthal coordinate (ϕ) of the muon trajectory.

The inner wheel formed by doublet modules is placed before the endcap toroidal magnet, while the big wheel consists of the seven layers (triplet module plus two doublet modules) as shown in Fig. 6.22 and measures the muon trajectory in the bending direction by toroidal magnet.

Fig. 6.22 Big wheel of TGC chamber. Reprinted under the Terms of Use from [12] ATLAS Experiment © 2006 CERN. All rights reserved. The diameter of the Big wheel is about 25 m

6.3.2.3 Monitored Drift Tube

Over most of the η range, a precise measurement of the track coordinates in the principal bending direction of the toroidal magnetic field is provided by monitored drift tubes (MDT) chambers. The MDT system achieves a sagitta accuracy of 60 μm, corresponding to the momentum resolution of about 10% at $p_T = 1$ TeV.

The basic element of the MDT is pressurised drift tube with a diameter of 29.970 mm, operating with Ar/CO_2 gas (93% : 7%) at 3 bar. The electrons resulting from ionisation are collected at the central tungsten-rhenium wire with a diameter of 50 μm at a potential of 3080 V as shown in left figure of Fig. 6.23. The average drift velocity of electrons is about 20.7 μm/ns and the maximum drift time is about 700 ns. Making use of the radius-to-drift time relation (r-t relation), the distance of a muon track passing through the tube from an anode wire can be measured as a drift circle. The shape of the r-t relation, which depends on parameters such as temperature, pressure, magnetic field distortions caused by the positive ions after ionisation, must be known with high accuracy in order to achieve better spatial resolution.

The mechanical structure of an MDT chamber is shown in right figure of Fig. 6.23. A chamber consists of two multi-layers of three or four drift tube layers. In order to monitor the internal geometry of the chamber, four optical alignment rays, two parallel and two diagonal, are equipped. That is why the drift tube detector in the ATLAS experiment is called "Monitored" Drift Tubes. The 1,150 MDT chambers are constructed from 354,000 tubes and cover an area of 5,500 m^2. Each MDT chamber provides the information of the track segment. Muon tracks are reconstructed by track segments obtained from inner, middle and outer stations of MDT chambers.

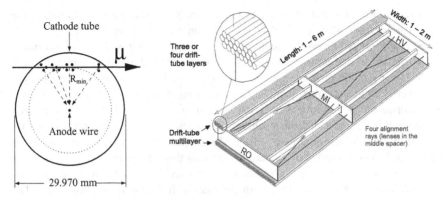

Fig. 6.23 Left: the cross-section of the MDT drift tube. Right: the mechanical structure of a MDT chamber. Reproduced by permission of IOP Publishing from [11] © IOP Publishing Ltd and SISSA.

6.3.3 Muon Reconstruction

The muon reconstruction can be performed independently in the inner tracker and MS. In the inner tracker, the muons are reconstructed such as any other charged particles described in Sect. 6.1. In this section, the description of the muon reconstruction in the MS and the combined muon reconstruction are focused on. More detail on the muon reconstruction at the ATLAS experiment is given in Ref. [13].

Using the drift circles in MDTs or clusters in TGCs and RPCs, the muon reconstruction is subdivided into the three stages: segment-finding, segment-combining and track-fitting.

Segment-finding starts with a search for hit patterns in a single station (i.e. inner, middle and outer stations of MDT, RPC and TGC chambers in case of the ATLAS MS) to form the track segments. The Hough transform is used to search for hits aligned on a trajectory in the detector. The track segments are reconstructed by a straight-line fit to the hits found in each layer.

Full-fledged track candidates are built from segments, typically starting from middle stations of detector where trigger hits from TGC or RPC are available, and extrapolating back through the magnetic field to the segments reconstructed in the inner stations. Whenever a match of the segment is found, the segment is added as the track candidate. The final track-fitting procedure takes into account all relevant effects: multiple scattering, non-uniformity of the magnetic field, inter-chamber misalignment etc.

The physics analyses make use of four muon types.

- Combined muon: muon tracks reconstructed by the inner tracker and MS independently are combined with a global refit using the hits from the inner tracker and MS detectors. In order to improve the fit quality, MS hits may be added to or removed from the track. Most muon tracks are reconstructed by outside-in reconstruction, where the muons are first reconstructed in the MS and then extrapolated inward and match to a track reconstructed by the inner tracker. An inside-out reconstruction where the reconstruction procedure is opposite to the outside-in reconstruction is also used as a complementary approach.

- Segment-tagged muons: a muon track in the inner tracker is classified as a muon if it is associated with at least one local segment in the MDT stations. In case of low p_T muon or in case muons pass through the $\eta - \phi$ region, which is not covered by MS stations, segment tagged muons are used.

- Calorimeter-tagged muons: a muon track in the inner tracker is identified as a muon if it is associated with an energy deposit in the calorimeter compatible with a minimum-ionising particle (MIP). Muons passing through the $\eta - \phi$ region where MS is not fully covered are regarded as this type of muons.

- Extrapolated muons: muon tracks reconstructed based only MS and a loose requirement on compatibility with originating from the interaction point. The tracking parameters of the muon are defined at the interaction point, taking into account the estimated energy loss of the muon in the calorimeters. Extrapolated muons are used to extend the acceptance for the muon reconstruction into the region where the inner tracker does not cover.

When the same track reconstructed by inner tracker is identified by two muon types, the priority is given to the combined muons, then to segment-tagged muons, and finally calorimeter tagged muons.

6.3.4   Muon Identification

Although muon candidates reconstructed by the muon spectrometers are mostly true muons, we want to identify the origin of muons. Muons from the decay of heavy particles such as W, Z, Higgs bosons, or new particles are interesting for us and need to be reconstructed as "isolated" muons efficiently and precisely. Since muons from semi-leptonic decays from b and c-hadrons and τ are also important for the b-tagging and τ ID, respectively, they need to be reconstructed as muons in the heavy flavour jets and τs. On the other hand, muons from the decays of pions and kaons are regarded as "fake" muons and eliminated from muon candidates.

Muon candidates originating from in-flight decays of charged hadrons mainly from pion and kaon decays in the inner tracker are reconstructed with a distinctive kink in the track. Therefore, it is expected that the track fit quality of the resulting combined track is poor and that the momentum measured by the MS and the inner tracker are not compatible. Muon identification is performed by applying quality requirements to suppress the background, to select prompt muons with high efficiency, and to guarantee a robust momentum measurement. Based on the number of hits in the inner tracker and MS, χ^2 of the combined muon tracks, the difference between the transverse momentum measurements in the inner tracker and MS and their uncertainties are used to classify as "Loose", "Medium", "Tight" and "High p_T" (for high momentum muons above 100 GeV aimed at the muons from exotic particle such as Z' and W' bosons) categories. These categories are provided to address the specific needs of different physics analyses.

6.3.5 Muon Isolation

Muons originating from the decay of heavy particles such as W, Z or Higgs bosons are often produced isolated from the other particles, in contrast to the muons from semi-leptonic hadron decays such as $b \to c\mu\nu$, which are embedded in jets. The measurement of the detector activity around a muon candidate, referred to as muon isolation, is a powerful tool for background rejection in many physics analyses. Both track-based and calorimeter-based isolation variables are often used.

The track-based isolation variable $p_T^{\text{varcone}30}$ is defined as the scalar sum of the transverse momentum of the tracks $p_T > 1$ GeV in a cone size $\Delta R = \min(10 \text{ GeV}/p_T^\mu, 0.3)$ around the muon. The muon momentum p_T^μ is excluded from $p_T^{\text{varcone}30}$. In this case, the cone size is chosen either to be p_T dependent ($\Delta R = 10 \text{ GeV}/p_T^\mu$) or to be p_T independent ($\Delta R = 0.3$). The p_T dependent cone size is used to improve the performance for the isolated muon with a high transverse momentum. The calorimeter-based isolation variables $E_T^{\text{topocone}20}$ are defined as the sum of the transverse energy of topological cluster in a cone size $\Delta R = 0.2$ around the muon. The isolation selection criteria are determined using the relative isolation variables defined as $p_T^{\text{varcone}30}/p_T^\mu$ and $E_T^{\text{topocone}20}/p_T^\mu$. Several selection criteria are provided to address the specific needs of different physics analyses.

6.3.6 Momentum Scale and Resolution

Although the simulation contains the description of the detector, there is a limitation in describing the momentum scale and the momentum resolution. For this reason, corrections of simulated values are often applied. The momentum scale and resolution are parameterised by the following equation:

$$p_T^{\text{Cor}} = \frac{p_T^{\text{MC}} + \sum_{n=0}^{1} s_n(\eta, \phi) \times \left(p_T^{\text{MC}}\right)^n}{1 + \sum_{m=0}^{2} \Delta r_m(\eta, \phi) \times \left(p_T^{\text{MC}}\right)^{m-1} g_m} \tag{6.4}$$

where p_T^{MC} is the uncorrected transverse momentum in simulation, g_m is normally distributed random variables with zero mean and unit width, and the $\Delta r_m(\eta, \phi)$ and $s_n(\eta, \phi)$ are the parameters representing the smearing of momentum resolution and the scale corrections applied in a specific (η, ϕ) detector region, respectively.

The corrections to the momentum resolution are described by the denominator of Eq. (6.4), assuming that the relative p_T resolution can be parameterised by

$$\frac{\sigma(p_T)}{p_T} = \frac{r_0}{p_T} \oplus r_1 \oplus r_2 \cdot p_T \tag{6.5}$$

with $\oplus$ denoting a sum in quadrature. As shown in Sect. 6.3.1, the second and third terms of Eq. (6.5) account mainly for multiple scattering and the resolution effects caused by spatial resolution of the hit measurements and the misalignment of the muon spectrometer. The first term accounts for fluctuation of the energy loss in the detector material. The difference in the momentum resolution between data and simulation is parameterised by $\Delta r_m(\eta, \phi)$. The momentum in simulation is smeared with the $\Delta r_m(\eta, \phi)$, by dividing uncorrected muon momentum by the term of denominator in Eq. (6.4).

The numerator in Eq. (6.4) describes the momentum scales. The $s_1(\eta, \phi)$ corrects for inaccuracy in the description of the magnetic field integral and the dimension of the detector in the direction perpendicular to the magnetic field. The $s_0(\eta, \phi)$ corrects the energy loss in the detector material.

The momentum scale and resolution are usually studied using $J/\psi \rightarrow \mu\mu$ and $Z \rightarrow \mu\mu$ decays. Since the J/ψ and Z are narrow resonances and their masses are well known, the distributions of invariant mass reconstructed by two μ's from J/ψ and Z show clear peaks around 3 GeV and 91 GeV [14], respectively. Furthermore, the number of non-resonant background events from decays of light and heavy hadrons and from continuum Drell-Yan production is very small. The momentum scale and resolution are determined from data using a fit with templates derived from simulation, which compares the invariant mass distributions from $J/\psi \rightarrow \mu\mu$ and $Z \rightarrow \mu\mu$ candidates in data and simulation. The momentum in the range of 5 GeV< p_T <20 GeV and 20 GeV< p_T <300 GeV is corrected by $J/\psi \rightarrow \mu\mu$ and $Z \rightarrow \mu\mu$ candidates, respectively. Figure 6.24 shows the invariant mass distribution of $J/\psi \rightarrow \mu\mu$ (left) and $Z \rightarrow \mu\mu$ (right) candidate events reconstructed with combined muons [13]. The agreement between data and simulation becomes much better after the correction.

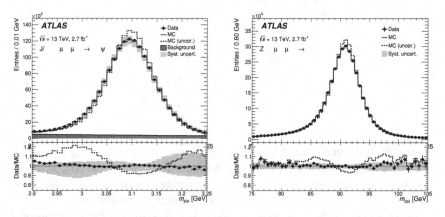

Fig. 6.24 Dimuon invariant mass distribution of $J/\psi \rightarrow \mu\mu$ (left) and $Z \rightarrow \mu\mu$ (right) candidate events reconstructed with combined muons. Reprinted under the Creative Commons Attribution 4.0 International License from [13] © CERN for the benefit of the ATLAS collaboration 2016. The upper panels show the invariant mass distribution for data and for the signal simulation, and for background estimate. The points show the data, the continuous line shows the simulation with the corrections of momentum scale and resolution, and the dashed lines show the simulation without the corrections

6.4   Jet Identification

6.4.1   Fragmentation: Partons to Particles

This section describes the identification and reconstruction of jets. A jet in high-energy physics is, naively speaking, a bunch of hadrons, which are emitted in nearby directions. This is an object consisting of the consequence of parton(s) fragmented into multi-hadron states. Here gives a short introduction of how we understand the "fragmentation" process, i.e. the underlying physics of the partons transformed into long-lived hadrons, followed by discussion on algorithms to identify and reconstruct jets.

The partons, i.e. quarks and gluons, obey the dynamics described by QCD with one and only parameter, the strong coupling constant α_S. The coupling constant becomes smaller with the energy of the interaction as a consequence of the renormalisation group equation as shown in Fig. 2.6. The energy scale, denoted as μ, is given as the centre-of-mass energy of the partons in concern. Since the energies involved in each parton reaction is not measurable, the choice of the energy scale for a process is, however, not uniquely given and we leave the discussion to elsewhere. Here, we merely point out that there are many choices: it could be centre-of-mass energy or transverse momentum of two partons when discussing on the parton-parton collisions, often quadratically summed with a heavy quark mass if a heavy quark is involved, or the mass of the particles (W, Z, Υ ...) when discussing the decay of particles.

Now let us take a simple example, a decay of the Z^0 boson into a $q\bar{q}$ pair for understanding how a parton fragment into a multi-hadron state. The energy scale would be given as $\mu = m_{Z^0}$. In this case, the quarks cannot be a pair of top quarks due to energy conservation and the quarks run fast and may radiate additional gluons, since a quark feel the force from the other quarks due to the colour charge carried by each of quarks. The force is "strong", as α_S is about 0.1 at the mass scale of m_{Z^0}. The radiation of the gluon is *soft* in most cases, i.e. typically collinear and/or with small momentum fraction with respect to the parent quark, like for the case of bremsstrahlung. But with a small probability, the gluon may have large angle from both of the two quarks *and* may have large momentum fraction, i.e. the radiated gluon may be *hard*.

The gluons and quarks still feel the colour force and may further radiate a pair of $q\bar{q}$, $g \rightarrow q\bar{q}$ or radiate further a gluon $g \rightarrow gg$. The splitting of partons would be repeated: this process is often called "parton shower". After some steps of radiation, the partons are branched into many parton states, most of which have another close-by parton and the invariant masses between these two partons are much smaller than the initial mass m_{Z^0}. In such a situation, the coupling constant describing the interaction of the partons becomes much larger, say 0.3 rather than 0.1. This accelerates the process of the fragmentation and eventually all the partons would have their invariant mass with the nearest partons below 1 GeV. This is the energy scale of Λ_{QCD}, below which perturbative QCD (pQCD) is no longer applicable; one cannot discuss the branching of partons by perturbation theory and need a help of the non-perturbative approach, e.g. the lattice QCD.

The lattice QCD calculation tells us that the potential energy of a $q\bar{q}$ pair is linear to the distance r between the pair, $U(r) \propto r$. This means that the line of the strong force is about constant density and concentrated in a tube-like area. The stored energy gets higher as the distance becomes larger. Once the stored energy exceeds beyond the mass of two quarks, the total energy should be lower if a $q\bar{q}$ pair is produced and the force lines are cut. This process continues until the relative distances between all the colour-neutral $q\bar{q}$ pair become shorter that there remains no more enough energy to produce additional $q\bar{q}$ pair, giving the end of the showering process. All the quarks and gluons are in bound states, i.e. mesons or baryons, at this stage. This last part of the transition is called hadronisation.

Prior to the discussion on jet algorithm, we may like to see if the input to the algorithm, the four momentum of the final state objects, is well defined. The final state with hadrons can clearly be defined once we give a threshold on the lifetime of the final state particles. The boundary is often given at where the B−mesons and charm mesons decay but not charged pions. The intermediate states of partons before hadronisation, on the other hand, are less obvious in their definition and we need certain criteria, which we discuss in the following section. One should also note that the fragmentation is a process described by quantum field theory and it is not possible to assign a certain parton or hadron to their parents in principle—what we know through the theory is the probability to which parents the daughter particle is assigned, unless the lifetime of the parent particle is long enough that the quantum effect is negligible.

6.4.2 Defining Jets

While it is impossible to have unique one-to-many correspondence between a parton and hadrons, one may still imagine that a spray of particles, or a jet, would be originated from a quark or a gluon, if it looks like collimated and away from the other activities, and may like to relate the jet to the underlying parton. This relies on the fact that the parton emission is mostly collinear if the parents of the hadronised partons have sufficiently high energy and run fast. In practice, however, the jets are still "ambiguous": if there are two close-by jets or a wide jet, it is often not straightforward to know whether to associate a parton or two or more partons to the jets. A hadron away from the sprays may also be ambiguous in such an assignment or left unassociated to any jets.

Things are even more complicated if we are to reconstruct jets using the detector information. Not only the momentum of the particles are smeared by the detector but also that a significant part of the particles may be escaped from detection. Also, a measurement by calorimetry cannot resolve two close-by hadrons since their energy clusters may be merged to a cluster if the distance between the two hadrons at the calorimeter is less than a certain value.

This also applies when we extend the concept of the jets to partons. Parton branching is also a quantum-mechanics process and the final state partons are not uniquely related to their parent partons, as described above.

These facts all call for some definition of jets, or a jet algorithm that defines the number of jets and their momenta. The algorithm has to be independent of the type of particles: a few "primary" partons, many partons after the parton shower, hadrons before or after the meson decays, or detector measurements. The algorithm should also identify, count and reconstruct the momentum of hard, i.e. high momentum partons while the soft emissions nearby hard partons should be absorbed to the hard partons, or discarded. In this sense, the algorithm should be insensitive to the soft emissions, or 'infrared safe'. In fact, the procedure to absorb the soft particles to the stronger jets is somewhat analogical to the procedure of renormalisation in theoretical calculation.

6.4.3 Jet Algorithms

Historically, there have been two kinds of jet algorithms used in high-energy physics, one called the cone algorithm and the other the cluster algorithm. The cone algorithm moves around a window of jet area defined as a circle in $\eta - \phi$ space, $\Delta r = \sqrt{\Delta \eta^2 + \Delta \phi^2} < R$, where the Δ variables Δr, $\Delta \eta$ and $\Delta \phi$ are the distance between the jet centre and the position of the particle. R is called cone radius, giving the angular boundary of jets. The algorithm iteratively finds such an energetic cluster where the sum of the transverse momenta p_T^{jet} for the particle inside the circle becomes maximum. The cluster with transverse momenta is said to be a jet if $p_T^{\text{jet}} > p_T^{\text{thr}}$, the threshold value of the jets, which is the parameter to ensure that the jets are hard. After the algorithm is run, one may find many jets but also many particle clusters below the p_T^{thr}, which are not qualified to be a jet and discarded. This feature is suitable for hadron-hadron collisions where hard jets are accompanied with many soft particles arising from soft emissions from beam remnants, particles from soft underlying events (rescattering of the outgoing proton remnants) and multi-parton interactions, those often called "underlying events". In addition, particles from soft collisions pile up to the hard partons in case of high-luminosity collisions such as the main LHC runs (see Sect. 6.4.4).

There still remains the activity of particles not emerged from the hard partons within the cone. Although the amount of the underlying events and pile-up particles are certainly not constant, it is at least possible to statistically subtract such contributions since the size of the jet is the same for each jet unless they overlap and some part of the jet area is to be shared; even for such a case the net area of overlapped jets is still well defined. This is another virtue of the cone algorithm.

The historical cluster algorithm, on the other hand, assigns all the particles to one of the jets. This is suitable for e^+e^- collisions where there are neither beam remnants, multi-parton interactions nor pile-up but only the soft emissions from hard partons. The basic idea is that the soft particles should always be merged to other particles or nearest cluster until the energy of the cluster exceeds beyond the threshold (see Fig. 6.25). The "distance" between two particles, d_{ij} is defined in various ways, which gives the variation and choice to the algorithm. The distance can be an invariant mass squared between two particles (the original JADE algorithm) or relative transverse

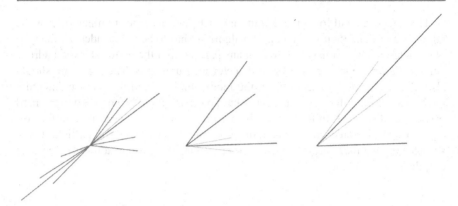

Fig. 6.25 A schematic drawing, showing how clusters with the nearest distance are merged each other to form a new cluster

momentum squared of a softer particle with respect to the harder particle (often called k_T or $k_\perp$): $d_{ij} = min(p_T^{i2}, p_T^{j2})$. The algorithm sequentially combines two particles of the nearest distance. In each step of the combination, the four momentum of the two merged particles is calculated to form a new particle. There are also many choices in how to combine the momenta of two particles (called "recombination scheme")—either the merged particle is massless, massive, conserving energy or not etc. The definition of distance and recombination scheme should be chosen such that the jet observables in concern (momentum, energy, number, mass etc.) are well reproduced, and the choice may vary with energy and type of the interaction. The combination is stopped until the distance between two jets, defined as $y = d_{ij}/M$ become above y_{cut}, where M is normally chosen as the invariant mass of the first two outgoing partons from the e^+e^- collisions, which equals to the centre-of-mass energy of the e^+e^- collisions in most of the cases, except for the events with hard initial state radiation of photons.

The biggest advantage of the cluster algorithm against the cone algorithm is that there is no ambiguity in the algorithm originated from the iterative procedure in the cone algorithms. The cone algorithm needs seeds to start the iteration. It is well known that the number of jets and jet momenta is largely affected by the choice of the property of the seed (p_T threshold and the cone size to define a seed) when particles are densely populated. A jet could be split into two jets depending on the seed choice. This means that the result of the jet finding is affected by soft particles (which could be the seed). It is known that a naive cone algorithm is not infrared safe, i.e. the result of the algorithm may depend on a presence of a particle with infinitesimally small energy.

A new class of cluster algorithms for hadron-hadron collisions are then invented, by taking virtue of the cone algorithms, (a) the algorithm works on $\eta - \phi - p_T$ space so that it is boost invariant and (b) particles below the threshold are discarded. A typical arrangement is to introduce two particles with infinite momentum on the beam axis. In the algorithm, the distance between the beam particles d_i is defined as $d_i = p_T^{i2}$, where p_T^i is the transverse momentum of the ith particle, in addition to k_T

Fig. 6.26 A schematic drawing, showing how clusters with large momenta are classified: to be merged each other to form a new cluster, or to the beam axis to be considered as a jet

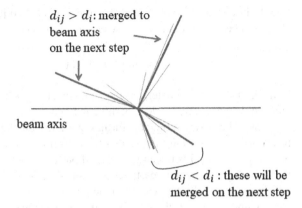

as the distance parameter between two final state particles, d_{ij}. If d_i of the particle, the distance to the beam axis, is smaller than any of d_{ij}'s, the distances to the other final state particles, the particle is merged to the beam particle.

Also, d_{ij} is adjusted to the hadron collider environment. The first version of the algorithm, the k_T algorithm, uses the distance parameter as

$$d_{ij} = \min(p_T^{i2}, p_T^{j2}) \Delta r_{ij}^2 / R^2 \qquad (6.6)$$

where Δr is that used in the cone algorithms, $\Delta r = \sqrt{\Delta \eta^2 + \Delta \phi^2}$, and R is the radius parameter. The particle with the smallest p_T will be merged to the beam if Δr is more than R for any other particles, since then d_i would be smaller than any of d_{ij}. It will be merged to the nearest particle if $\Delta r < R$. In this way, the parameter R plays the role of the cone radius in cone algorithms (Fig. 6.26).

The value of R gives the angular size of the jets. This is a parameter to which extent one allows to include hard parton radiation around the primary parton, in addition to the soft emissions and/or collinear part of radiated partons. The size should not be too small to include the soft/collinear particles but should not be too large since the particles from beam-related activities (soft underlying events, multi-parton interactions and pile-up particles) may come more into the jet area. Typical values used for the QCD studies at the energy scale of weak interactions ($p_T^{jet} \simeq m_Z/2$) is $0.6 - 0.7$ to include soft emission originated from the parent partons, in order to reduce the theoretical uncertainty in pQCD description of the data. For higher energy interactions, $0.4 - 0.5$ would be more preferred, in particular for physics beyond the SM (BSM) searches at TeV scale, to minimise the effect of soft particles to the momentum or mass reconstruction of the parent BSM particles.

The clustering procedure is finished when there is no possibility to merge the remaining particles each other except for the beam particle. The particles above a given p_T threshold is defined as jets and others are discarded, i.e. merged to the beam particle.

The original k_T algorithm, where d_{ij} is defined as Eq. (6.6), it is known that the jet area tends to be extended beyond the area given by the parameter R. This feature is undesirable for the mass reconstruction as discussed just above. Recently *anti-k*$_T$ algorithm became more popular, where d_{ij} is defined as:

$$d_{ij} = \min((p_T^i)^{-2}, (p_T^j)^{-2}) \Delta r_{ij}^2 / R^2$$

With this distance parameter, the algorithm first merges the pairs within the maximum allowed distance, $\Delta r_{ij} \simeq R$, making a merged particle in between. The same thing happens in the next iteration: the furthest particle from the new merged particle would be absorbed. This would imply that the direction of the jet particle, or the jet axis, would be oscillated between the merged particles, but the axis will be stabilised in the later stage where only particles with small k_T with respect to the jet axis are left, which are eventually be merged to the jet. As a consequence, the jet area will have clear boundary of a circle with the radius R. The area of overlapping circles will be absorbed to more energetic jets. This would give the jets very similar to what is given by the cone algorithm, which has certain area size. One can statistically subtract the underlying events of the jets in such a case. The anti-k_T algorithm combines virtues of the cone and cluster algorithm successfully and is now the most popular jet algorithm at the LHC.

Naturally, the criteria to define the jets of the partons is based on continuous parameters, such as p_T, R, which have no characteristic scale, apart from Λ_{QCD}, being anyhow much below the typical jet momentum $(> O(10)\,\text{GeV})$. There are some arbitrariness on the parameter values in such algorithms, and the parameters may have to be optimised for each application.

6.4.4 Calibrating Jet Measurements

As described in Sect. 6.4.1, the jets are expected to be collimated at high energies, primarily since radiation in the final state is less pronounced for high-p_T jets where the coupling constant α_S is smaller. There the individual hadrons consisting of a jet cannot easily be resolved, and the calibration of the detector is performed at the level of jets instead of constituting hadrons. The result of jet calibration is often called jet energy scale (JES).

Since a jet is defined by means of an algorithm, the momenta of jets depend on the choice of the algorithm and jet finder parameters (e.g. R). The calibration on jets should be repeated for each choice of the algorithm and the set of parameters. There is another point to choose: if the energy is to be corrected to the "particle level", i.e. jets using the momentum of particles, or to the parton level, where partons are the input for the jet algorithm. A general consensus is that the correction to the particle level is to be applied, i.e. the momentum of the particle-level jet gives the reference to a detector-level jet, which matches in $\eta - \phi$ space. In this way, one can avoid the theoretical uncertainty on the correction factor from the particle to the parton level jets, which is expected to be improved as the theory is more advanced.

A simplest reconstruction of jets is to start from only calorimeter information, as described in Sect. 6.4.3. The energy calibration for calorimeter objects is done to the electromagnetic scale, ignoring that $e/h \neq 1$ (see Sect. 5.4.3), or to the hadronic scale after applying e/h correction. In principle, a calorimeter cluster can be replaced with a matched track to improve the momentum resolution, if the spatial resolution of the calorimeter cluster is fine enough to resolve close-by particles.

There remains still difference between the particle-level and detector-level jets. In addition to the factors arising from calorimetry, such as longitudinal hadron shower leakage and intrinsic dependence of calorimeter response on type of particles (the e/h ratio, response for muons and neutrinos), the following effects specific to jets cause significant shift in measured momentum:

- particles escaping outside the jet area

  If the jet area is wider than the radius used in the jet finder, the particles outside the jet is lost and the jet energy is underestimated. As explained above, the size of the jet is wider for low energy jets since the partons are radiated more often. This leak, however, should be a part of the jet definition for both parton and particle-level jets. Further leak occurs when the particles are bent by solenoidal magnetic field applied to the central tracker. This is to be corrected through the jet energy scale.

  A gluon radiates another partons more often than a quark because of the different colour factor (9/4 vs. 1). In general, a gluon jet is wider than a quark jet and particle spectrum is softer because of more radiation. A gluon jet contains more particles with smaller average energy than for a quark jet. This again leads to more leaks by the magnetic bending of particles.

  An additional correction depending on the jet properties, such as the transverse size of the jet or the number of tracks matched to the jet, would improve the energy resolution of jets.

- response of heavy-quark jets

  Some shift may remain for c-quark and b-quark jets (c/b-jets). A b-jet may decay semi-leptonically, to a lepton (e, μ, τ), a neutrino and a lighter c- or u-quark jet. The c-quark jet may decay again semi-leptonically. As a consequence, c/b-jets may contain one or more electron or muon and one or more neutrinos. The momentum of neutrinos cannot be measured; moreover, the muon leaves only up to two GeV energies in calorimeter (MIP). Therefore, the energy responses for b- and c-quark jets are, in general, smaller than other kinds jets (u, d, s or gluon jets—often called "light flavour jets"). The actual difference depends, then, on many factors, e.g. on how the muon momentum is taken into account, the e/h ratio, which may affect the jet energy containing an electron etc. Anyhow, it is a common practice to apply additional correction if a jet is identified as a heavy quark jet.

- pile-up

  One can safely assume that the events that pile up on top of a collision of interest are all soft interactions (see Sect. 2.5). The soft interaction events are often called "minimum-bias" events since, the events are taken through triggers as little requirement as possible, e.g. small energy in very forward part of the calorimeter. Average p_T from such minimum-bias events at the LHC energy ($\sqrt{s} \simeq 14\,\text{TeV}$) is about 2.4 GeV per unit of $\eta - \phi$ space. This means that a jet with radius $R = 0.4$ with 20 additional minimum-bias events have average offset of p_T by about 24 GeV, hence gives a very large shift in energies for jets with $p_{Tjets} < O(100)$ GeV. In order to reduce the influence from pile-up particles, the expected average p_T from pile-up is subtracted. The actual value to be subtracted depends on the number of pile-up events. The average number of pile-up can be estimated from the luminosity of the collisions. The Poisson fluctuation from the average can further be corrected by measuring the number of interactions per bunch crossing through, e.g. N_{PV}, the number of primary vertex per crossing reconstructed from the central tracker.

The residual difference from imperfect simulation of the detector and remaining miscalibration of detectors is corrected by in-situ measurements of jet response. The most common way to determine the overall jet energy scale is to find and use some physics processes with a jet whose energy can be deduced through energy-momentum conservation. In the hadron collider experiments, for example, the production of γ+jet or Z+jet is widely used as the calibration source, where the photon or Z reconstructed from dilepton can be the reference to the jet energy because the fluctuation of energy deposited by the electromagnetic shower or measured charged track momentum is much smaller than that by the hadronic shower, leading to more precise energy measurement than that by the hadron calorimeter. However, since the momentum conservation is hold only in the plane perpendicular to beam axis in the hadron collider, what is conserved is p_T, not p. More concretely, in γ+jet event, the jet energy scale is adjusted so that the p_T of the jet is equal to that of the photon. The result of the calibration is illustrated in Fig. 6.27. The clear peak can be seen in the γ+jet events with the peak close to unity as expected.

With the similar concept of calibrating the electromagnetic scale, $Z \to q\bar{q}$ can be used in principle as the calibration source of the jets with the Z mass as the reference target. However, this method does not work in the hadron collider experiments in practice, because of the overwhelming dijet backgrounds generated by QCD process.

Fig. 6.27 The ratio of p_T of jet to γ for the p_T range between 160 and 210 GeV. Reprinted under the Creative Commons Attribution 4.0 International License from [15] © CERN for the benefit of the ATLAS collaboration 2014. The calorimeter region is restricted to be $|\eta| < 1.2$

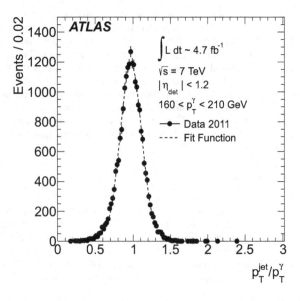

In addition, the jet energy resolution is much worse than that of the electromagnetic energy measurement, resulting in the difficulty to see the resonant peak from $Z \rightarrow q\bar{q}$.

6.5   Reconstructing Missing Momentum

The momentum of the neutral particles, such as neutrino and unknown neutral particles, are not detected by collider detectors. For hadron colliders, the longitudinal momentum of such particles cannot be known due to the lack of longitudinal momentum information of the collisions. The missing transverse momentum, denoted as either p_{Tmiss} or $\mathbf{E}_{\mathrm{Tmiss}}$, can still be reconstructed by the negative of the transverse momentum vector of observed particles. A first approximation of the sum of the visible particle momentum could simply be obtained from the x and y component of the calorimeter cell energies, $E_{i,\mathrm{cell}} \sin \theta_i \cos \phi_i$ and $E_{i,\mathrm{cell}} \sin \theta_i \sin \phi_i$. This would miss muon momenta and also detailed calibration depending on the final state objects are ignored. Instead, one may measure each category of final state objects separately, with proper calibration and possibly with a help of tracking and muon detectors, for example,

$$\mathbf{p}_{\mathrm{Tmiss}} = \sum \left[\mathbf{p}_{\mathrm{T}}^e + \mathbf{p}_{\mathrm{T}}^\mu + \mathbf{p}_{\mathrm{T}}^\gamma + \mathbf{p}_{\mathrm{T}}^\tau + \mathbf{p}_{\mathrm{T}}^{\mathrm{jets}} + \mathbf{p}_{\mathrm{T}}^{\mathrm{others}} \right],$$

as is done for the ATLAS experiment. Here $p_{\mathrm{T}}^{\mathrm{others}}$ term is the momentum of the particles belonging to neither of objects identified as charged lepton, jet nor a photon. This term, often called as "soft term", includes the rest of the particles accompanied with the hard interaction, such as particles from ISR, multi-parton events and underlying events, which should also be added to the p_{Tmiss} calculation. The soft term, however, also includes particles from pile up. Since average transverse energy of a minimum-bias event is about 100 GeV, the total transverse energy of an event with >20 pile-ups would be about 2 TeV and increases proportionally as a function of the number of pile-up. The resolution of this pile-up component directly affects the missing p_{T} calculation. Therefore, the performance of the missing p_{T} reconstruction strongly depends on how to estimate the missing vector from the soft term, and to less extent through the jet term. In addition, any misreconstruction in the detector, such as noise in the calorimeter, affects to $\mathbf{p}_{\mathrm{Tmiss}}$ through the soft term.

Various algorithms are developed in order to mitigate the growth of the resolution with the number of pile-up events. A simple algorithm is to reconstruct the soft term by using only the calorimeters or the central tracker. The latter has a benefit that it can remove all the track momentum not originated from the vertex of the hard scattering in concern. It misses, however, the contribution from the neutral particles like π^0 and K_L^0. A further refined algorithm could be to reweight the calorimeter soft term by the momentum fraction of the soft term from trackers from the primary vertex (PV): $\Sigma_{\mathrm{tracks,PV}} p_{\mathrm{T}} / \Sigma_{\mathrm{tracks}}$. Each estimator would have different resolution and tail. Now suppose that we find long tail in for zero missing-p_{T} events on truth level. It is found there that tails of the soft term of tracking and calorimeter algorithms are not strongly correlated. For that reason, it is often useful to use more than one algorithm to reduce the tail induced by the p_{Tmiss} reconstruction. This also indicates that the choice of the soft term reconstruction depends on the type of events in concern.

6.6   Identification of b-Jet and τ-Jet

In many types of physics analyses in the collider experiments, the identification of b-quark jet (b-jet) or τ-jet is of particular importance. For example, the Higgs boson has the large branching fractions of $H \rightarrow b\bar{b}$ and $H \rightarrow \tau^{+}\tau^{-}$. The top quark decays into b and W with almost 100% of probability. This section describes the methods to identify b-jet and τ-jet.

6.6.1 b-Jet

There are two approaches for the b-jet identification, or b-tagging. The first one exploits the fact that b-hadron generated at the collision point travels a few mm before the decay ($c\tau$ of B^{0} is 455 μm for example), leaving the secondary vertex or the collection of tracks that have large values of the impact parameter with respect to the primary vertex. We refer this type of b-tagging as the track-based tagging below. The second one exploits the fact that the b-hadron decay is associated with leptons with high probability because the branching fraction of the semi-leptonic decay of b-quark is approximately 11%. In addition, b-quark decays to c-quark plus something with the probability close to 100%, where the semi-leptonic branching fraction of c-quark is about 10%. Hence, the existence of a lepton nearby a jet can be a signature of b- or actually also c-quark jet. We refer this second type of b-tagging as the soft lepton tagging. In either method, all jets are the candidate of b-jets, i.e. all jets are examined if they are originated from b-quarks. In the actual application, two methods are often combined. Or more precisely, there are some branches in the track based tagging, and the discriminants from each tagging method are often unified with a multivariate analysis technique for better discrimination.

6.6.1.1   Track Based Tagging

The track-based b-tagging makes use of the difference in lifetime between b-hadrons and other more common particles generated by the collisions, such as pions or protons. As schematically shown in Fig. 6.28, b-hadrons fly typically a few mm from the collision point before the decays, hence producing particles that emerge from the space point away from the primary vertex. On the other hand, light quarks, such as u, d, or gluon, generate only light hadrons that appear from the beam-beam interaction point, causing charged particles associated with the primary vertex. Using this difference, the track-based b-tagging algorithms search for either tracks with their impact parameter significantly away from zero, or explicitly reconstruct secondary vertex formed by the decay products of b-hadrons. One thing the reader should keep in mind is the existence of particles generated at the primary vertex even in the b-jet,

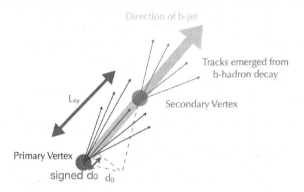

Fig. 6.28 Schematic drawing of b-jet. d_0 is the impact parameter of the tracks with respect to the primary vertex. The signed impact parameter is defined to be the distance of the d_0 projected to the jet axis with a sign, where the sign is defined to be positive if the track crosses the jet axis in the region towards jet direction in a view from the primary vertex, and negative if the crossing point is behind the primary vertex. The decay length, L_{xy}, is defined to be b-hadron flight length or more specifically the distance between the primary and the secondary vertices in x-y plane

because of the bi-products of b-quark hadronisation, which is also shown in Fig. 6.28. These particles sometimes degrade the b-tagging capability because they mimic jets originated from light quarks. This becomes more striking if the momentum or energy of the original b-quark is higher. More energetic partons end up with more particles through hadronisation, while the number of decay products of b-hadron does not depend on momentum of parent b-hadron or b-quark, i.e. the fraction of particles from the primary vertex compared to the one from the secondary vertex increases as the b-quark gets harder.

Figure 6.29 shows the signed impact parameter significance of the tracks in the simulated $t\bar{t}$ events. The definition of the signed impact parameter is explained in the figure caption of Fig. 6.28. As can be seen, b-jets have more tracks with the large value compared to the other jets originated from u-, d-, s-quark, or gluon, which are referred to as light jets. By the way, the perfect detector that has the infinite position resolution would not give us the negative value of the signed impact parameter because the vector drawn from the primary to the secondary vertex should be the same as the jet direction. Therefore, with such detector, if existed, the signed impact parameter distribution has monochromatic peak at zero plus the tail to only the positive side due to the contribution from b- or c-hadrons etc. In other words, the negative value is caused by the detector resolution. The width of the peak around zero, therefore, represents the detector resolution.

The most simple application in the track-based tagging is to just count the number of tracks with some selection criteria to pick up the tracks that do not come from the primary vertex. In slightly more complicated approaches, the likelihood of the track impact parameters is formed and used as the discriminant. One example is the jet probability algorithm, where the probability density function of the impact parameter for the tracks that come from the primary vertex is created a priori, and then the likelihood value of each charged particle to be consistent with the one from the primary vertex is calculated. In many cases, there are many tracks in a jet, and

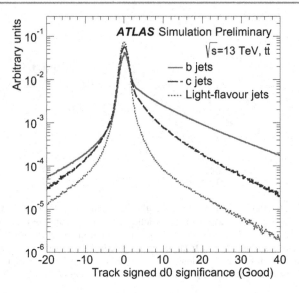

Fig. 6.29 Impact parameter significance distributions of tracks inside *b*-, *c*-, or light jets. The distributions are obtained in the ATLAS group simulation. Reprinted under the Creative Commons Attribution 4.0 International License from [16] © 2015 CERN for the benefit of the ATLAS Collaboration

hence the likelihoods assigned to each track are combined to form a likelihood or discriminant for a jet in concern. The benefit of this method is that one examines if a track is compatible with the hypothesis that it comes from a collision point, and no priori knowledge is required for *b*-jets.

In further application, one can construct the *b*-jet likelihood based on the probability density function for the tracks produced by the *b*-hadron decays a priori, for example, by the simulation. Taking the likelihood ratio for the *b*-jet hypothesis and the light-jet hypothesis would give us the improved discrimination power over the jet probability where only the light-jet hypothesis is used in principle. In the actual application, special care needs to be taken to form the *b*-jet probability density function. We must use a correct probability density function and would like to confirm that a priori knowledge or simulated data reproduces real data, but it is not so easy to extract non-biased *b*-jet sample with high purity. This is in contrast to handling the jet probability tagging where the light-jet sample can be as easily accumulated with high purity. Hence, the jet probability method is more robust, although the discriminating power is less and suitable for the usage in the early stage of the experiment.

Another type of track-based *b*-tagging algorithm explicitly reconstructs the secondary vertex caused by the decay of *b*-hadron. The secondary vertex is reconstructed as already described in Sect. 6.1.4. After finding the secondary vertex, the *b*-tagging algorithm usually set a threshold on the significance of the decay distance defined in Fig. 6.28, which is the distance from the primary and secondary vertices, or the flight length of the *b*-hadron in the plane perpendicular to beam axis.

Once the secondary vertex is formed, some extra information, such as the invariant mass calculated from the tracks associated to the secondary vertex etc., can be extracted, resulting in high rejection power for the light jets than the impact parameter based b-tagging method. However, the efficiency is the key issue, because one needs at least two tracks in the secondary vertex finding, while even one track can give us the discriminating power at some degree in the impact parameter based tagging.

In either types of algorithms, the impact parameter based or the secondary vertex reconstruction, one also needs to remove the tracks which emerge from the secondary vertex although their origin is not b-hadron. Tracks generated by the decay of K_S and Λ, and by photon conversion are the typical example. In many applications, the algorithm looks for two-track combination whose invariant mass is consistent with K_S, Λ and photon and removes them in the track list to consider.

6.6.1.2 Soft Lepton Tagging

A b-jet contains charged lepton nearby with high probability, which comes either from direct semi-leptonic decay of a b-hadron or from the cascade decay through c-hadron. Another possible source of charged leptons is the leptonic decays of W or Z. But they don't produce additional jets, i.e. such leptons are "isolated". This difference, isolated vs. non-isolated, is very frequently and efficiently used to discriminate whether the charged lepton in question is originated from the b-jet or W/Z. The typical application to identify b-jet is therefore to require a jet to have a charged lepton nearby, for example, ΔR between the jet and charged lepton is required to be smaller than a threshold. This technique is called as soft lepton tagging.

In principle, we can use any charged leptons for the soft lepton tagging. However, only muons are used in practice because of the difficulty in identifying τ's and non-isolated electrons. The identification of τ-jet is discussed in the next section. The non-isolated electron often shares its electromagnetic shower with the shower or energy deposit by the constitutes of the jet. This is in contrast to the muon case. It's only muon that can penetrate the calorimeter and reach the muon detector even with jets. Hence, non-isolated muons can be still identified with high efficiency and low fake rate.

The possible background source of the soft lepton tagging with muon is either the punch through (see Sect. 3.3.2) or decay of hadrons. This is basically the background in the muon identification. Thus, the muon identification capability mostly determines the performance of the soft muon tagging.

To achieve higher b-jet selection efficiency or suppress the fake contribution from light jets, a kinematical requirement for the non-isolated muon is sometimes imposed. Suppose we know the direction of the jet axis. This axis is a good approximation of the initial b-quark momentum vector, or flight direction of the b-hadron, which is produced by the hadronisation of the initial b-quark. In this process, the non-isolated lepton momentum transverse to the b-hadron flight direction or approximately the jet axis can be as large as a half of the b-hadron mass. On the other hand, there is no mechanism for hadrons yielded from light quarks to get the momentum transverse to the jet axis rather than the tiny contribution in the hadronisation process. Thus, the transverse momentum relative to the jet axis can give us some discriminating power between b-jets and other types of jets.

6.6.2 τ-Jet

The τ identification is classified into few categories based on that it decays either leptonically or hadronically, and the number of final states particles. The branching fraction of leptonic decay is about 35%. In the remaining 65% cases, the τ decays hadronically with one charged particles (one prong) with the fraction of about 50%, and with three charged particle (three prong) with the fraction of 15%.

In case of leptonic decays, the τ identification is actually the identification of isolated electron or muon as the final state consisting of either electron or muon and neutrino which is not detected. In absence of another neutrino, the sum of momentum vector of the isolated electron or muon, and missing E_T can be treated as the momentum of τ. This is rather straightforward and cleaner method compared to the identification in hadronic decays.

The identification for hadronic decays is more complex, but important because of the larger branching fraction. In the hadronic decay, there are one or three charged particles often associated with extra neutral particles such as π^0. Since we are now dealing with rather high momentum τ's, the decay products are boosted and collimated, resulting in a τ-jet. The particles inside the τ-jet are basically decay products of the τ, hence the number of particles, which does not depend on τ momentum, is typically smaller than the one in quark or gluon induced jet (hadronic jet) where the number of particles depends on the momentum. This leads to the fact that the particles or the energy carried by the particles inside the τ-jet are more collimated than that in the hadronic jets, given the same jet energy. In addition, the hadronisation process could generate a particle whose momentum relative to the jet axis is greater than the half of τ's mass due to QCD radiation. On the other hand, the maximum in τ decay is a half of τ mass. This is another reason why the τ-jet is more collimated. In terms of the width of shower shape, the other important point we have to care is electromagnetic shower, which mimics τ shower. As we have seen in the previous sections and chapters, the size or width of electromagnetic shower is smaller than the one of hadronic shower. This means that electron could be the fake of τ, if you just select collimated jet. Therefore, we have to require the jet width to be narrower than hadronic jet and wider than electromagnetic shower at the same time.

Another feature of τ-jet is that τ has a finite lifetime, whose $c\tau = 87\,\mu$m, causing a possible decay vertex in addition to the primary vertex created by collisions. This means that the track-based b-tagging can also give us the discrimination of τ-jet from the hadronic jets in principle. However, this lifetime is much shorter than that of b-hadrons. Therefore, using a method similar to track-based b-tagging alone does not produce sufficient discriminating power. Still the track information helps to identify τ in cooperation with the jet shower shape variables.

The actual τ-jet identification starts from reconstructing a jet. Usually, no special jet clustering algorithm for τ-jet is used. The similar one for the hadronic jet is used

Fig. 6.30 One of the
discriminating variables used
in the *τ* ID in the ATLAS
group, which is the fraction
of calorimeter energy in the
region $\Delta R < 0.1$ to the total
energy in a jet. Reprinted
under the Creative Commons
Attribution 4.0 International
License from [17] © 2011
CERN for the benefit of the
ATLAS Collaboration. The
red histogram shows the
distribution for *τ*'s in
$Z \to \tau\tau$ or $W \to \tau\nu$
simulated events, while the
black dots for inclusive jets
in real data

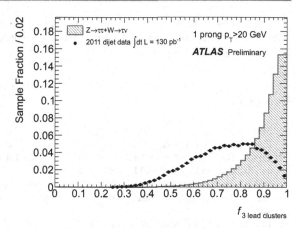

Fig. 6.31 Discriminating
variables used in the *τ* ID in
the ATLAS group, which is
the maximum of ΔR
between the tracks inside a
jet and the jet axis. Reprinted
under the Creative Commons
Attribution 4.0 International
License from [17] © 2011
CERN for the benefit of the
ATLAS Collaboration. The
red histogram shows the
distribution for *τ*'s in
$Z \to \tau\tau$ or $W \to \tau\nu$
simulated events, while the
black dots for inclusive jets
in real data

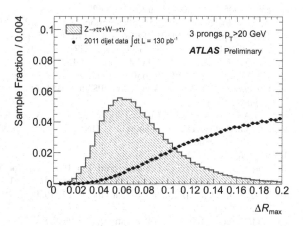

with the parameters possibly tuned for *τ*-jet clustering. The jet considering here is
a cluster based on energy deposit in the calorimeter. The next step is to select and
associate the tracks to the *τ* candidate jet. Some quality cuts and the requirement on
p_T are the standard criteria. If necessary the *τ*-jet candidate is categorised into one
or three prong based on the number of associated tracks.

Here, we show you some variables that are actually used in the ATLAS *τ*-jet
identification. Figure 6.30 shows the fraction of calorimeter energy in the region
$\Delta R < 0.1$ to the total energy in the jet for $Z \to \tau\tau$ or $W \to \tau\nu$ MC and for real
data where most of the jets originate from light-quarks or gluons. Figure 6.31 shows
the maximum of ΔR between the tracks inside a jet and *τ*-jet axis. As can be seen
from these two figures, the energy flow of the *τ*-jet is concentrated on the centre of
the jet.

The τ identification algorithm nowadays exploits the multivariate analysis such as likelihood, neural network or boosted decision tree based on the variables discussed above or some other variations.

References

1. Fruhwirth, R., Regler, M., Bock, R.K., Grote, T., Notz, D.: Data Analysis Techniques for High-Energy Physics. Cambridge University Press (first published in 1990, second edition in 2000)
2. Cowan, G.: Statistical Data Analysis. Oxford University Press (1998). Roe, B.P.: Probability and Statistics in Experimental Physics. Springer (1992)
3. https://twiki.cern.ch/twiki/bin/view/AtlasPublic/EventDisplayRun2Physics
4. https://atlas.web.cern.ch/Atlas/GROUPS/PHYSICS/CONFNOTES/ATLAS-CONF-2011-161/
5. Lampl, W., Laplace, S., Lelas, D., Loch, P., Ma, H., Menke, S., Rajagopalan, S., Rousseau, D., Snyder, S., Unal, G.: ATL-LARG-PUB-2008-002
6. Aad, G., et al.: [ATLAS Collaboration]. Eur. Phys. J. C **77**, 490 (2017). https://doi.org/10.1140/epjc/s10052-017-5004-5. arXiv:1603.02934 [hep-ex]
7. Aaboud, M., et al.: [ATLAS Collaboration]. Eur. Phys. J. C **79**(8), 639 (2019). https://doi.org/10.1140/epjc/s10052-019-7140-6. arXiv:1902.04655 [physics.ins-det]. https://atlas.web.cern.ch/Atlas/GROUPS/PHYSICS/PAPERS/PERF-2017-01/
8. Aaboud, M., et al.: [ATLAS Collaboration]. Eur. Phys. J. C **79**(3), 205 (2019). https://doi.org/10.1140/epjc/s10052-019-6650-6. arXiv:1810.05087 [hep-ex]
9. https://atlas.web.cern.ch/Atlas/GROUPS/PHYSICS/EGAMMA/PublicPlots/20100721/display-photons/
10. ATLAS Collaboration, CERN-LHCC-97-022 ; ATLAS-TDR-10
11. Aad, G., et al.: [ATLAS Collaboration]. JINST **3**, S08003 (2008). https://doi.org/10.1088/1748-0221/3/08/S08003
12. https://cds.cern.ch/record/986163
13. Aad, G., et al.: [ATLAS Collaboration]. Eur. Phys. J. C **76**(5), 292 (2016). https://doi.org/10.1140/epjc/s10052-016-4120-y. arXiv:1603.05598 [hep-ex]
14. Zyla, P.A., et al.: [Particle Data Group]. PTEP **2020**(8), 083C01 (2020). https://doi.org/10.1093/ptep/ptaa104
15. Aad, G., et al.: [ATLAS Collaboration]. Eur. Phys. J. C **75**, 17 (2015). https://doi.org/10.1140/epjc/s10052-014-3190-y. arXiv:1406.0076 [hep-ex]
16. ATLAS Collaboration, ATL-PHYS-PUB-2015-022
17. ATLAS Collaboration, ATLAS-CONF-2011-152

Event Simulation

7

Not only real experimental data but also simulated data are necessary for modern data analyses because experiments, that is, detectors, etc. are getting complex so that we need the help of Monte Carlo (MC) simulation to understand the experimental data. A *Monte Carlo* method is a technique to simulate high-energy physics interactions and detector responses by applying random samplings to probability distribution modelling experiments. In collider experiments, MC simulated events (MC events or MC samples in short) are produced event-by-event; in the case of proton-proton colliders, each event corresponds to a bunch crossing, where several pp collisions (pile-up) may occur. MC simulation is also useful to design new experiments and detectors.

7.1 Overview

We outline the production of MC events with MC simulation, where three steps are considered: event generation, detector simulation and reconstruction as shown in Fig. 7.1.

In the event generation step, we produce events, for example, two photons from Higgs bosons in proton-proton collisions ($pp \rightarrow gg \rightarrow H \rightarrow \gamma\gamma$), or two quarks from Z bosons in electron-positron collisions ($e^+e^- \rightarrow Z \rightarrow q\bar{q}$). Unstable particles, whose lifetime is short enough not to reach detectors, are decayed according to branching fractions which are obtained from experimental measurements or theoretical predictions. The output of this step is a list of particles with various information, for instance, energy, momentum, production and decay positions, status and relation between particles, i.e., parent and children.

In the detector simulation step, we simulate our detector responses to the stable particles produced in the previous step. For example, in the case of electrons, they lose energy by interacting with detectors: produce electron-hole pairs in pixel and silicon

© The Author(s) 2022

K. Hanagaki et al., *Experimental Techniques in Modern High-Energy Physics*,
Lecture Notes in Physics 1001, https://doi.org/10.1007/978-4-431-56931-2_7

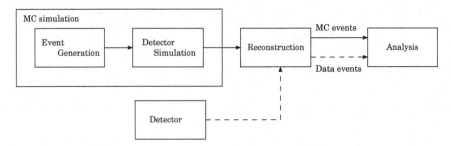

Fig. 7.1 The flow of the production of MC events (*full* detector simulation) and real data for data analysis. The *Monte Carlo* method is used in the event generation and detector simulation steps but not in the reconstruction step. The reconstruction step should work for both MC events (solid) and real data (dashed)

detectors, ionise particles in gas detectors and produce EM showers in calorimeters. These energy deposits are converted to charges if necessary. The format of outputs is the same as that of the real data from detectors because the next step should be applied to both real data and MC simulation. The detector simulation should ideally be as precise as possible but it depends on the requirements of physics achievements and the technical limitations, for example, modelling of detectors, computing resources, etc. The pile-up effect in the pp collisions can be taken into account after the detector simulation, for example, we prepare several events of the inelastic interactions and mix them following the number of collisions per bunch crossing.

In the reconstruction step, we reconstruct events from the output of the detector simulation and identify particles. Reconstructed objects in each event are still *candidates* of particles. For instance, electron *candidates* mean that they are reconstructed from EM calorimeter clusters matched with charged tracks. They come from electrons (called true electrons) or fake electrons ($\pi^{\pm}$, τ, etc.); note that the fraction of true electrons is not so high. Then, identification programs are applied in order to select, for example, true electrons from electron *candidates* as much as possible. In other words, fake electrons are rejected as much as possible. As a result, the fraction of true electrons in the selected electron *candidates* becomes high. Momentum and energy are also calculated for each object including calibrations if possible. The output of this step is used in the data analysis.

7.2   Event Generation

An MC event is produced with several steps, where each step uses different programs. We explain the outline of how to produce an event using a concrete example of the $t\bar{t}$ process in pp colliders with several keywords often used for MC production: matrix-element event generator, parton density function, parton shower, fragmentation, harmonisation, underlying event, etc. The detail of theoretical aspects can be found in books, for example, [1]. Then, we give concrete computing programs used in the ATLAS experiment.

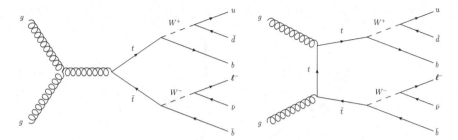

Fig. 7.2 Feynman diagrams of $t\bar{t}$ process in the pp collisions: $gg \rightarrow t\bar{t}, t \rightarrow W^+b, W^+ \rightarrow u\bar{d}, \bar{t} \rightarrow W^-\bar{b}, W^- \rightarrow \ell^-\bar{\nu}$ with s-channel (left) and t-channel (right)

7.2.1 Production of $t\bar{t}$ Process

Let us consider the production of $t\bar{t}$ events in pp collisions: $t\bar{t}, t \rightarrow W^+b, \bar{t} \rightarrow W^-\bar{b}$ where one of W bosons decay into quarks and the other into leptons. Figure 7.2 shows two Feynman diagrams of this production: $gg \rightarrow t\bar{t}, t \rightarrow W^+b, W^+ \rightarrow u\bar{d}, \bar{t} \rightarrow W^-\bar{b}, W^- \rightarrow \ell^-\bar{\nu}$. This is so-called a hard-process part. Matrix-element event generators (ME generators) are used for the production of the hard-process part. ME generators can produce events by considering all the diagrams which have the same final state. ME generators also perform the simulation of a gluon (g) from a proton in the proton-proton collision. The momentum fraction of gluons is described by parton density functions (PDF), which are obtained based on QCD and experimental measurements. Gluon and quark PDFs depend on the energy of the interaction and we need to define an energy scale to evaluate PDFs, for example, as top mass for $t\bar{t}$ production. This scale is one of the important parameters in the MC production, which is called a factorisation scale μ_F.

However, the production of a hard-process using ME generators is not the end of the story but a starting point of the event generation. To produce MC events, several different steps (parton shower, fragmentation/hadronisation, etc.) have to be performed as shown in Fig. 7.3. Lots of quarks and gluons with relatively small p_T are emitted using a method of parton shower, that is, soft and collinear emissions. A Sudakov form factor, which is a probability not to emit a parton until a target energy scale, is calculated to perform the MC method. There are two different types of showers for such emissions: initial state radiation (ISR) and final state radiation (FSR); the theoretical idea behind them is similar, however, ISR is complicated due to so-called backward evolution, which is "tracing the showers backwards in time" (see, for example, [2]). As shown in Fig. 7.3, the use of ISR and FSR depends on when the emissions happen; ISR (FSR) is for radiations before (after) the hard-process. ISR gluons and quarks are radiated from the initial gluons of $gg \rightarrow t\bar{t}$ but FSR gluons are radiated from quarks (u and/or $\bar{d}$) of $W^+ \rightarrow u\bar{d}$.

Quarks and gluons cannot be observed due to the colour confinement (see Sect. 6.4), so that they need to be combined to compose hadrons. This procedure is called hadronisation. For example, to produce a B meson ($B^0(\bar{b}d), B^+(\bar{b}u)$), u or d quarks are produced from parton shower/fragmentation and one of them is

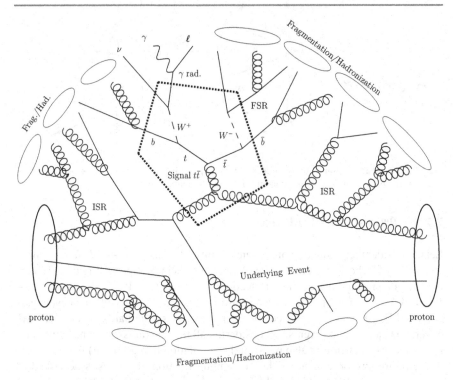

Fig. 7.3 Details of MC event production of $t\bar{t}$ from the proton-proton collisions: parton show-ers (ISR, FSR), fragmentation, hadronisation, photon radiation and underlying event. The hard-process shown with a dashed box is the same as the left plot of Fig. 7.2. Redrawn from Fig. 1 of [3]

combined with a $\bar{b}$ quark. New additional quarks and gluons are created from the existing quarks, gluons or vacuums, which are gluon fields in this case. This step is called fragmentation but is sometimes included in the step of hadronisation. Photons can be radiated from charged leptons and quarks, which are called QED radiative corrections. The remaining parts of protons not used in the hard-process are called "underlying event" and must be treated properly with parton showers, hadronisa-tion and fragmentation. Finally, the decay of mesons, baryons and leptons is per-formed until particles produced from decays become "stable". Kinematics (energy and momentum) of all the particles are determined under conservation rules (energy, momentum, spin/polarisation, etc.) with the MC method.

Feynman diagrams shown in Fig. 7.2 are the leading order (LO) of $gg \rightarrow t\bar{t}$ process and there is no additional gluons and quarks. However, we can use ME generators to produce more high p_T gluons and quarks. For example, when we consider one additional strong coupling, additional gluon or quark can be emitted, which should be treated by the ME generators instead of the parton shower. This is a part of the contribution of next-to-leading order (NLO). Gluons or quarks produced by ME generators and by parton shower are properly treated to avoid a double counting, which is briefly explained in the next section.

7.2.2 Event Generators

Many event generators are available on the market. Several main generators used in ATLAS for Run 1 and Run 2 data analysis are listed in alphabetical order: ALPGEN [4], HERWIG [5–7], MADGRAPH (MADGRAPH5_AMC@NLO) [8], MC@NLO [9], PHOTOS [10], POWHEG [11], PYTHIA [2,12], SHERPA [3], TAUOLA [13], etc. Since different generators have different features, each generator has its own pros and cons.

HERWIG, PYTHIA and SHERPA are multi-purpose event generators. They can do all the steps explained before including parton showers, fragmentation, hadronisation and decay. Not only LO but also NLO calculations for ME are available for a part of processes in these generators. However, other ME generators like MADGRAPH, POWHEG, etc. are often used for NLO in ATLAS. HERWIG and PYTHIA can be used for simulating parton showers, fragmentation, hadronisation and decay in these ME event generators. The theoretical idea of the parton shower is different among generators: angular ordering for HERWIG and p_T (or k_T) ordering for PYTHIA. Models for fragmentation and hadronisation are also different: cluster model for HERWIG and string model for PYTHIA. Their difference is often used as systematic uncertainties from the parton shower and fragmentation/hadronisation models. HERWIG has three major versions; HERWIG, HERWIG++ and HERWIG 7. PYTHIA has two major versions: PYTHIA 6 and PYTHIA 8. HERWIG++, HERWIG 7 and PYTHIA 8 were often used in ATLAS in Run 2 compared to HERWIG and PYTHIA 6 since new features and techniques of event generations were only implemented in HERWIG++, HERWIG 7 and PYTHIA 8.

ALPGEN, MADGRAPH and SHERPA are multi-leg generators. They can produce events with ME including multi-partons. The multi-partons are so-called additional partons or additional jets, so that we often call such physics processes X+jets, for example, W+jets, Z+jets, etc. The idea behind such additional jets with ME (ME jets) is that the modelling of jets produced with parton showers (PS) might not work well because the parton shower is based on soft and collinear approximation. Figure 7.4 shows one of the diagrams for W+2-jets and these additional quark and gluon associated with a W boson can be produced by either ME or PS. Calculations based on ME is in principle correct (see also the next paragraph, however); when the existence and behaviour of additional jets are critical in data analysis, the use of ME generators is recommended to describe additional jets, in particular, high p_T jets, but we should be aware that the implementation of loop corrections, etc. may depend on each generator. For example, in the SUSY searches, typical SUSY events from gluinos and squarks have several jets in the final state (e.g., $gg \rightarrow \tilde{g}\tilde{g} \rightarrow qq\tilde{\chi}_1^0 qq\tilde{\chi}_1^0$), and one of dominant background processes is $Z \rightarrow \nu\bar{\nu}$+jets. The "+jets" of the Z process must be modelled well to predict background, so that SHERPA is used for Z plus up to 4-jets with ME in ATLAS [14].

There is one important thing to be considered, which is called a jet-parton matching. Even if multi-leg generators are used to produce additional jets with ME, the parton shower has to be applied to produce additional jets, in particular, low p_T jets. Practically we assume that ME takes care of high p_T jets and PS does low p_T jets because ME jets are well modelled in high p_T region. To ensure this

Fig. 7.4 Feynman diagram
of W+jets: 2 jets (a quark
and a gluon) are associated
with a W boson

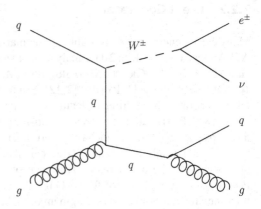

assumption, reconstructed high p_T jets must match to partons produced by ME. In
addition, jets in some phase spaces are produced by both ME and PS, which leads
to double counting of events. For this purpose, MLM prescription [15], CKKW-
matching procedure [16,17], etc. are applied and events are discarded if they cannot
satisfy their requirement. To explain MLM prescription in ALPGEN, let us consider
the production of W+up to 2-jets with a ME generator and jets with $p_T > 20$ GeV
are used for data analysis. First, events are produced with W with 0 ME-jet, W with 1
ME-jet and W with 2 ME-jets, separately, where additional partons with, for example, $p_T > 15$ GeV are produced by a ME generator. Note that the parton shower
is also applied, so that some high p_T jets might be produced by PS. Then, all the
reconstructed jets with $p_T > 15$ GeV are checked if they matched to ME-parton and
we count such jets. For W with 0 (1) ME-jet, we require such jets should be exactly
0 (1). For W with 2 ME-jets, we require such jets should be 2 or more. Then, we
merge the remaining events to make a W+up to 2-jets events.

MADGRAPH, MC@NLO, POWHEG and SHERPA are used as NLO generators for
some specific processes. Additional one parton and also loop diagrams up to the
next-to-leading order are properly taken into account.

PHOTOS generates QED radiative corrections for charged leptons and quarks.
TAUOLA is a program to simulate tau-decay including polarisation properly. They
are optionally used in PYTHIA and HERWIG.[1]

7.2.2.1 Cross Section

MC events are produced by using event generators, which can provide their cross
sections including branching fractions. However, in many cases, we don't use cross
sections provided by event generators but values obtained from dedicated programs,
because such programs can perform more higher order calculations than event generators. NLO event generators are available for most important physics processes on

---

[1] The recent version of PYTHIA 8, for example, 8.2 can treat tau-decay polarisation properly without
TAUOLA. We need to check the updates to use any generators.

Table 7.1 Cross section measurements at the ATLAS experiment ($\sqrt{s} = 13$ TeV) with theoretical predictions. The measurements are given with statistical and systematic uncertainties. The predictions are given with combined uncertainties including PDF, α_S, scales, etc. For the Higgs production cross section, five processes are included: gluon fusion (NNNLO), VBF (approximate NNLO), VH (NNLO/NLO), $t\bar{t}H + tH$ (NLO) and $b\bar{b}H$ (NNLO/NLO)

Process	Measurement	Prediction (higher order for QCD)	References
$W \to \ell\nu$ ($\ell = e, \mu$)	$20.64 \pm 0.02 \pm 0.70$ nb	$20.08^{+0.65}_{-0.66}$ nb (NNLO)	[18]
$Z \to \ell\ell$ ($\ell = e, \mu$)	$1969 \pm 1 \pm 56$ pb	1886^{+51}_{-57} pb (NNLO)	[19]
$t\bar{t}$	$826.4 \pm 3.6 \pm 19.6$ pb	832^{+40}_{-45} pb (NNLO+NNLL)	[20]
Higgs	$55.4 \pm 3.1^{+3.0}_{-2.8}$ pb	55.6 ± 2.5 pb (NNNLO, etc.)	[21]
ZZ	$17.3 \pm 0.6 \pm 0.8$ pb	$16.9^{+0.6}_{-0.5}$ pb (NNLO)	[22]
$t\bar{t}W$	$0.87 \pm 0.13 \pm 0.14$ pb	0.60 ± 0.07 pb (NLO)	[23]

the marker but NNLO event generators are limited. On the other hand, the dedicated programs can calculate cross sections up to NNLO (or higher for some processes).

There is an important parameter, i.e., a renormalisation scale μ_R, which is the scale at which the strong coupling is evaluated in order to calculate cross sections. The value of cross sections does not depend on the choice of μ_R; however, since we cannot perform complete calculations including all the orders of the strong coupling, the calculated cross sections might depend on μ_R. In the high-energy region ($\gg O(100 \text{ MeV})$), a perturbative method works well in QCD like QED and we can calculate cross sections, for instance, up to next-to-next-to-next-to-leading order (NNNLO, N3LO or N^{3}LO) for the Higgs gluon fusion production.

Including the higher order calculations, cross sections can be properly predicted and they are well consistent with the experimental results. Table 7.1 shows the results of $W, Z, t\bar{t}$, Higgs, ZZ and $t\bar{t}W$ production cross section measurements as concrete examples with theoretical predictions.

7.3 Detector Simulation

Particles produced in the event generators are detected through the interactions with several detector components (materials). To reach the detector volume including the beam pipe, particles has a long enough lifetime, that is, they are "stable particles" from a viewpoint of the detector simulation. Such stable particles are electrons ($e^{\pm}$), photons, $\pi^{\pm}$, kaons ($K^{\pm}$, K^0_S, K^0_L), $\mu^{\pm}$, protons ($p/\bar{p}$), neutrons and neutrinos in case of the SM. In addition, in the SUSY and other new physics models beyond the SM, some particles, for example, the lightest neutralino is stable in the SUSY models with R-parity conservation. Some of the stable particles, for example, $\pi^{\pm}$, $K^{\pm}$, K^0_S and $\mu^{\pm}$ can decay according to their lifetime in the detector volume, which is done in the detector simulation step, not by the event generator.

GEANT4 program [24] is a detector simulation toolkit widely used in the experimental particle physics. We build each detector component and define its interactions based on its real detector in order to emulate how particles interact and how much energy of particles is lost; energy loss and multiple scattering (charged tracks in tracking volumes); electromagnetic interaction/shower; hadronic interaction/shower; etc. This is based on the best knowledge of the particle interaction with materials. GEANT4 traces particles step-by-step and simulates their interaction. This is a reason why MC simulation with GEANT4 is called *full* simulation.

There are different types of MC simulations: "fast simulation" and "parametric simulation".[2] In the parametric simulation, the detector response is described by expected resolution functions for each stable particle. Momentum and energy are smeared with the resolution functions. The effects of reconstruction and identification programs, that is, their efficiencies are replaced with weights or MC methods following their expected performance. "Fast simulation" is sometimes the same as the parametric simulation but this term is also used in the case that a part of the detector simulation step, for example, calorimeter response, is replaced with a faster algorithm to emulate detector response. In terms of the modelling of the real data, in general, the full simulation is better than the fast and parametric simulations. However, from the point of view of execution time, the full simulation is much slower; for example, in some extreme cases, several minutes per event with the full simulation but less than a few seconds with the parametric simulation. If we need lots of events to reduce the MC statistical uncertainties, the use of fast or parametric simulations is one of the options. In addition, it takes a much longer time to develop computing programs with GEANT4.

References

1. Ellis, R.K., Stirling, W.J., Webber, B.R.: Camb. Monogr. Part. Phys. Nucl. Phys. Cosmol. **8**, 1–435 (1996)
2. Sjostrand, T., Mrenna, S., Skands, P.Z.: JHEP **0605**, 026 (2006). https://doi.org/10.1088/1126-6708/2006/05/026, arXiv:hep-ph/0603175
3. Gleisberg, T., Hoeche, S., Krauss, F., Schonherr, M., Schumann, S., Siegert, F., Winter, J.: JHEP **0902**, 007 (2009). https://doi.org/10.1088/1126-6708/2009/02/007, arXiv:0811.4622 [hep-ph]
4. Mangano, M.L., Moretti, M., Piccinini, F., Pittau, R., Polosa, A.D.: JHEP **0307**, 001 (2003). https://doi.org/10.1088/1126-6708/2003/07/001, arXiv:hep-ph/0206293
5. Bahr, M., et al.: Eur. Phys. J. C **58**, 639 (2008). https://doi.org/10.1140/epjc/s10052-008-0798-9, arXiv:0803.0883 [hep-ph]
6. Bellm, J., et al.: Eur. Phys. J. C **76**(4), 196 (2016). https://doi.org/10.1140/epjc/s10052-016-4018-8, arXiv:1512.01178 [hep-ph]
7. Corcella, G., Knowles, I.G., Marchesini, G., Moretti, S., Odagiri, K., Richardson, P., Seymour, M.H., Webber, B.R.: JHEP **0101**, 010 (2001). https://doi.org/10.1088/1126-6708/2001/01/010, arXiv:hep-ph/0011363

---

[2] The terminology to specify the type of MC simulation is not unique but depends on experiments. The machine learning techniques are also used for the MC simulation, for example, GAN.

8. Alwall, J., et al.: JHEP **1407**, 079 (2014). https://doi.org/10.1007/JHEP07(2014)079, arXiv:1405.0301 [hep-ph]
9. Frixione, S., Webber, B.R.: JHEP **0206**, 029 (2002). https://doi.org/10.1088/1126-6708/2002/06/029, arXiv:hep-ph/0204244
10. Golonka, P., Was, Z.: Eur. Phys. J. C **45**, 97 (2006). https://doi.org/10.1140/epjc/s2005-02396-4, arXiv:hep-ph/0506026
11. Nason, P.: JHEP **0411**, 040 (2004). https://doi.org/10.1088/1126-6708/2004/11/040, arXiv:hep-ph/0409146
12. Sjostrand, T., Mrenna, S., Skands, P.Z.: Comput. Phys. Commun. **178**, 852 (2008). https://doi.org/10.1016/j.cpc.2008.01.036, arXiv:0710.3820 [hep-ph]
13. Jadach, S., Kuhn, J.H., Was, Z.: Comput. Phys. Commun. **64**, 275 (1991). https://doi.org/10.1016/0010-4655(91)90038-M
14. ATLAS Collaboration, ATL-PHYS-PUB-2017-006
15. Mangano, M.L., Moretti, M., Pittau, R.: Nucl. Phys. B **632**, 343 (2002). https://doi.org/10.1016/S0550-3213(02)00249-3, arXiv:hep-ph/0108069
16. Hoeche, S., Krauss, F., Schumann, S., Siegert, F.: JHEP **0905**, 053 (2009). https://doi.org/10.1088/1126-6708/2009/05/053, arXiv:0903.1219 [hep-ph]
17. Catani, S., Krauss, F., Kuhn, R., Webber, B.R.: JHEP **0111**, 063 (2001). https://doi.org/10.1088/1126-6708/2001/11/063, arXiv:hep-ph/0109231
18. Aad, G., et al. [ATLAS Collaboration]: Phys. Lett. B **759**, 601–621 (2016). https://doi.org/10.1016/j.physletb.2016.06.023, arXiv:1603.09222 [hep-ex]
19. Aaboud, M., et al. [ATLAS Collaboration]: JHEP **02**, 117 (2017). https://doi.org/10.1007/JHEP02(2017)117, arXiv:1612.03636 [hep-ex]
20. Aad, G., et al. [ATLAS Collaboration]: Eur. Phys. J. C **80**(6), 528 (2020). https://doi.org/10.1140/epjc/s10052-020-7907-9, arXiv:1910.08819 [hep-ex]
21. ATLAS Collaboration, ATLAS-CONF-2019-032
22. Aaboud, M., et al. [ATLAS Collaboration]: Phys. Rev. D **97**(3), 032005 (2018). https://doi.org/10.1103/PhysRevD.97.032005, arXiv:1709.07703 [hep-ex]
23. Aaboud, M., et al. [ATLAS Collaboration]: Phys. Rev. D **99**(7), 072009 (2019). https://doi.org/10.1103/PhysRevD.99.072009, arXiv:1901.03584 [hep-ex]
24. Agostinelli, S., et al. [GEANT4 Collaboration]: Nucl. Instrum. Methods A **506**, 250 (2003). https://doi.org/10.1016/S0168-9002(03)01368-8

Examples of Physics Analysis

8

In this chapter, we present the analysis of Higgs and new physics searches as examples of data analysis. The data handled here are already calibrated and the particle identification for each object is also done.[1] In the so-called "data analysis" of the collider experiments, the event selection, background estimation, and signal extraction or measurement including evaluating systematic uncertainties are performed.

8.1 Higgs

8.1.1 Higgs Production Mechanism in Hadron Colliders

There are some different processes of Higgs production. Figure 8.1 shows the Higgs production cross sections in pp collisions as a function of Higgs mass. The largest contribution comes from the gluon fusion (Fig. 8.2a), in which there is no additional topology or feature other than the Higgs production. Hence, the inclusive analysis (see Sect. 2.3) is enforced as long as we consider the gluon fusion process. On the other hand, the final states of the other three processes contain not only Higgs but also extra particles, resulting in the characteristic topologies.

The second-largest cross section is via vector boson fusion (VBF) process where either Ws or Zs radiated from quarks couple together producing a Higgs boson (Fig. 8.2b). The quarks radiating W or Z bosons appear as forward jets, because their p_T which tends to be close to the W or Z mass, is much smaller than the momentum of colliding protons. In addition, since this process does not contain any colour

---

[1] An object might be possible to be a different type of particle, for example, an electron or a tau. The final particle identification, that is, the assignment of a particle type to each object depends on data analysis.

© The Author(s) 2022
K. Hanagaki et al., *Experimental Techniques in Modern High-Energy Physics*,
Lecture Notes in Physics 1001, https://doi.org/10.1007/978-4-431-56931-2_8

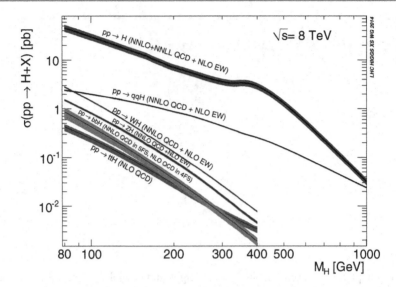

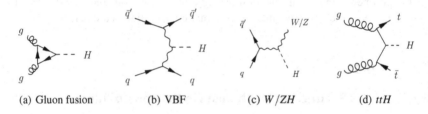

(a) Gluon fusion   (b) VBF   (c) W/ZH (d) ttH

Fig. 8.2 Feynman diagrams of Higgs productions

exchanges between the incoming quarks, no parton radiation would exist around the produced Higgs or the detector central region, in contrast to the overwhelming multijet background where not only hard jets but also many soft jets are produced. Putting what is mentioned so far together, the Higgs production through the vector boson fusion process has a very unique topology with two forward jets and with little QCD activities (partons due to colour exchanges) in the central region except for Higgs decays. The feature allows us to significantly reduce the background due to multijet productions as well as the other types of background.

Another important production mechanism is the associate production with a vector boson, i.e., either W or Z (Fig. 8.2c). In case the W or Z decays hadronically, it does not help to improve the signal-to-noise ratio due to the overwhelming multijet backgrounds. However, leptonic decays of W or Z produce isolated leptons, allowing us to significantly improve the signal-to-noise ratio with a cost of the small branching fractions of W and Z.

The production cross section of associate production of $t\bar{t}$ (Fig. 8.2d) is one order of magnitude smaller than that of WH production. It is still accessible because of the characteristic topology. This production mechanism has special importance because this allows the direct access to the top Yukawa coupling.

Below we describe the basic idea of the analysis for $H \rightarrow \gamma\gamma$, $H \rightarrow b\bar{b}$, and $H \rightarrow W^+W^-$.

8.1.2 $H \rightarrow \gamma\gamma$

The Higgs boson was discovered in the ATLAS [2] and CMS [3] experiments in 2012. In this discovery, $H \rightarrow \gamma\gamma$ and $H \rightarrow ZZ^* \rightarrow \ell\ell\ell'\ell'$ channels played the most important role because they can reconstruct the invariant mass of the Higgs boson precisely compared to other channels, for example, $H \rightarrow WW^* \rightarrow \ell\nu\ell'\nu'$ even if the expected statistics for $H \rightarrow \gamma\gamma$ and $H \rightarrow ZZ^* \rightarrow \ell\ell\ell'\ell'$ is not high. In the distribution of the invariant mass of the Higgs boson candidates, we can observe a clear peak of the signal on top of the background events, which is one of the most reliable evidence of a resonance particle to claim its discovery. In this section, we explain how to search for the Higgs boson with the $H \rightarrow \gamma\gamma$ channel in the ATLAS experiment.

As mentioned before, the signal statistics is limited since the branching ratio of $H \rightarrow \gamma\gamma$ is very small, about 0.2%, for the mass of around 125 GeV, while thanks to a good resolution of diphoton invariant mass $m_{\gamma\gamma}$, a narrow resonance was expected to be observed on a huge but smooth background as shown in Fig. 8.3 [4]. Below we'll explain how to obtain this result.

We need two photons to reconstruct the invariant mass of diphotons, which is a final discriminant to extract the signal. Events having two photon candidates must be recorded in the offline storage to perform the analysis and diphoton triggers (35 and 25 GeV for photon E_T) were used for the trigger selection. Since events with jets faking photons, which are called fake photons, are not negligible, we cannot use, for example, single-photon triggers with a low E_T threshold like 25 GeV.[2] In the analysis, two photon candidates were selected with $p_T > 40$ GeV and 30 GeV, which are high enough to ensure the offline selected events achieve 100% trigger efficiency. This is a common technique in the physics analysis, because the estimation of trigger efficiency is not easy in general, especially for the momentum close to the turn-on of efficiency. We can avoid using such events near the trigger turn-on by requiring much higher p_T in offline selection compared to the trigger level. In this way, the source of possible large systematic uncertainty can be removed with the cost of losing some fraction of signal events.

There are three different processes in the background events: two real photons, one real photon+one fake photon, and two fake photons, which are called $\gamma\gamma$, γ+jet, and dijet, respectively. These background events do not make a peak but a smooth falling curve in the diphoton invariant mass $m_{\gamma\gamma}$ distribution as shown in Fig. 8.3. These

---

[2] 120 GeV or higher is required to use single-photon triggers, which is much higher for the photons coming from the Higgs boson of 125 GeV mass.

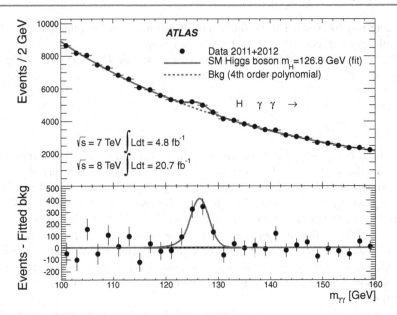

Fig. 8.3 Invariant mass distribution of diphotons for the combined 7 and 8 TeV data in ATLAS. Reprinted under the Creative Commons Attribution License 3.0 from [4] Copyright © 2013 CERN. The result of a fit to the data with the sum of a SM Higgs boson (126.8 GeV) and background is superimposed. The lower panel shows the residuals of the data with respect to the fitted background

compositions can be measured using photon identification variables, for example, an isolation variable. Their fractions were determined to be ∼74% for $\gamma\gamma$, ∼22% for γ+jet, and ∼3% for dijet. In addition, the Drell-Yan process ($Z^{(*)}/\gamma \to e^+e^-$, DY) remains with ∼1% of the background due to hard-bremsstrahlung.

It is important to improve the resolution of the $m_{\gamma\gamma}$ distribution. For this purpose, we need to measure photon energy and also the angle between two photons as precisely as possible. Since the EM calorimeter has three layers longitudinally in ATLAS, the direction of photons can be determined from the measurements of photon cluster positions. The production vertex of diphotons is calculated from the direction of two photons. This method is called *calo-pointing*. The position obtained with the calo-pointing is precise enough in terms of the $m_{\gamma\gamma}$ resolution while a more precise determination is required for the association of charged tracks to jets because jets from pile-up are identified using this association information. The production vertex position is finally obtained by using several information, for example, charged tracks not matched to any photons, charged tracks from conversions, the balance between two photons and charged tracks, etc. The resolution of the $m_{\gamma\gamma}$ is about 3%, and events with two unconverted photons have better resolution than those with at least one converted photon about 10% in relative.

Selected events are classified into several categories for two reasons; the first reason is to improve sensitivities for the search itself, which is called a global search here, and the second one is to measure properties of specific production processes, for

Fig. 8.4 Observed local p_0 as a function of the Higgs boson mass m_H for 7 TeV data (blue), 8 TeV data (red), and their combination (black). Reprinted under the Creative Commons Attribution License 3.0 from [4] Copyright © 2013 CERN. The dashed curves show the expected median local p_0 for the SM Higgs boson hypothesis when tested at a given m_H

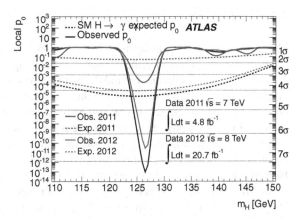

example, VBF and VH processes using extra leptons, jets, and the missing E_T. For example, 14 categories were introduced in the 8 TeV data analysis using jets, leptons, and transverse missing energy, where 2 for VBF, 3 for VH, and the other 9 categories for the improvement of the discovery sensitivity. There was about 30% improvement in the global search sensitivity compared to the result without categorisation.

The event excess, which is a signature of Higgs decays, is evaluated with a local p_0, which is a probability of how similar an observed distribution is to that with a background-only hypothesis. If the p_0 value is 0.5, it indicates that the observation is consistent with the background-only hypothesis, that is, no excess. If the p_0 value is smaller (larger) than 0.5, it means there is an excess (a deficit)[3] over the background. In addition, if a search is performed for a new narrow resonance ($\sim$4 MeV in case of SM Higgs boson) with an *unknown* mass in the invariant mass distribution ($m_{\gamma\gamma} = [110, 160]$ GeV in case of SM Higgs boson), we need to take into account the so-called look-elsewhere-effect. This effect can properly treat the fact that excesses like 3σ due to the statistical fluctuation could happen even if there is no new resonance in the search region and the frequency of such fake excesses becomes high in case of narrow resonance searches.[4] The p_0 value after taking this effect is called a global p_0.[5] This effect is negligible in the case of broad resonance searches due to an intrinsic particle width, worse detector resolutions, etc. With the full dataset of LHC Run 1 in ATLAS (2011–2012), the largest excess with respect to the background-only hypothesis (based on local p_0) was observed (expected) with 7.4 (4.3)σ at 126.5 GeV as shown in Fig. 8.4 [4].

---

[3] For the Higgs search in ATLAS, $p_0 = 0.00135$ (2.85×10^{-7}) corresponds to 3(5)σ, which is based on a one-sided limit.

[4] For example, we can assume that resonances with either 4 and 40 GeV width could exist in the mass range of 110−150 GeV. In this case, we may see more statistical fluctuations for 4 GeV signal than 40 GeV because the overall behaviour of the 40 GeV signal is not changed in the search range.

[5] In Ref. [2], the global significance of a local 5.9σ excess is estimated to be about 5.1σ in the mass range of 110–600 GeV. This result includes $H \to \gamma\gamma$, $H \to ZZ^* \to \ell\ell\ell'\ell'$, and $H \to WW^* \to \ell\nu\ell'\nu'$.

8.1.3 $H \rightarrow b\bar{b}$

This section outlines the analysis of $H \rightarrow b\bar{b}$. As the branching fraction of $H \rightarrow b\bar{b}$ is the largest ($\sim$58%) among the various decays of Higgs of 125 GeV mass, $H \rightarrow b\bar{b}$ could be the most useful and natural decay mode to search for Higgs and to study its properties in view of the statistics. On the other hand, the signature of the final state consists of just two b-jets. There would be no issues in the case of the e^+e^- colliders such as ILC, which provide a very clean environment experimentally, resulting in a very high signal-to-noise ratio. In the hadron colliders, however, the study of $H \rightarrow b\bar{b}$ is not straightforward at all because of the overwhelming QCD backgrounds (multijet background processes). At the energy of LHC, for example, the production cross section of inclusive b-jets is larger by the eighth order of magnitude than that of the Higgs. In addition, the identification of b-jets is not perfect. Light jets can mimic the signal. In this case, any jet production can be a background, whose production cross section is even higher than the inclusive b-jet cross section. Therefore, at the hadron colliders, we need some clever ideas to separate the $H \rightarrow b\bar{b}$ signals from the huge background.

In the following, we discuss the analysis method of $H \rightarrow b\bar{b}$ using the vector boson fusion process first and then the associate production of W and Z.

8.1.3.1 Vector Boson Fusion Process

The final state consists of two b-jets decayed from Higgs and two forward jets. Since there are no isolated leptons or large missing E_T which are commonly used to trigger an event, careful study and the optimisation of the trigger are needed. The most apparent choice of the trigger would be to require four jets with relatively high p_T. In addition a requirement on the topology, i.e., the existence of two forward jets in different η, respectively, may be applied if such a topological trigger is available. Even with the requirements above, still the remaining events would be dominated by the multijet background because of the huge production cross section. In order to suppress the multijet events further, the existence of a muon (see Sect. 6.6.1.2) that arises from the semi-leptonic decay of b-hadrons (directly or through the cascade decay to c like $b \rightarrow c\ell^-\bar{\nu}$) may be required with a cost of statistics. Even though there are two b-hadrons (and hence two c-hadrons followed by the decay of b-hadrons most of the time), the branching fraction of semi-leptonic decay is only the order of 10% (see Sect. 6.6.1) . The p_T of the lepton from the semi-leptonic decay is not so large. Because of these two factors, the signal efficiency is relatively low. Therefore, one has to optimise the trigger condition with a careful study. In other words, this is where the improvement potentially exists.

The offline analysis starts by selecting events with four jets. Out of the four, two are required to be in the central (rather small $|\eta|$), and the other two in the forward region ($\equiv$ forward jets). The forward jets tend to keep the direction of the parents' protons, and hence to be in the opposite region in η In order to select only the VBF process, commonly used requirements for the forward jets are to have a large separation in η between the two, where if one is in $\eta > 0$ then the other must be in $\eta < 0$, and to have large invariant mass reconstructed from the two forward jets.

Once an event passes the selection criteria for the forward jets, the remaining part is rather straightforward. The two central jets must be identified as b-jets, where there is always a room of the optimisation or tuning of the b-tagging requirement. For example, a requirement of at least one b-tag is also possible. The tightness of b-tagging requirement is another knob for tuning. Finally, we look for a signal peak in dijet mass distribution, which is reconstructed from the two central jets.

8.1.3.2 Associate Production with W or Z

The idea behind using the associate production with W/Z is to exploit an isolated lepton from W/Z decay to reduce background. In both trigger and offline event selection, an event is required to have at least one isolated lepton with some criteria such as p_T or η. Then in the offline selection, W can be identified by reconstructing transverse mass from the isolated lepton and the missing E_T. In the case of Z, dilepton mass is a powerful tool to separate the signal out from backgrounds.

The procedure after selecting or tagging W/Z is very similar to that in the VBF analysis. The dijet mass reconstructed from b-tagged jets is the most efficient variables to discriminate signal from background. In the end, the dominant source of backgrounds is W/Z production associated with heavy flavour jets, whose final state is exactly the same as the signal. On top of that, $t\bar{t}$ production is also a main component of the remaining background. Therefore, jet energy resolution to identify a possible peak from $H \to b\bar{b}$ decay is one of the most important key elements in this analysis, as well as the efficiency to detect and identify the final state objects. Figure 8.5 shows the distribution of dijet mass reconstructed from two b-tagged jets in the ATLAS experiment, where all the expected background contribution, except for $VZ, Z \to b\bar{b}$ ($V = Z$ or W), is subtracted. One can see a peak by $Z \to b\bar{b}$ as well as the small enhancement around 125 GeV, which is the evidence of $H \to b\bar{b}$.

Fig. 8.5 The invariant mass distribution reconstructed from two jets. Reprinted under the Creative Commons Attribution 4.0 International License from [5] © CERN for the benefit of the ATLAS collaboration 2021. The dots represent data. The red histogram shows the expected signal contribution where the signal yield is assumed to be 1.06 times the standard model expectation. The grey histogram shows the expected background contribution by ZZ or WZ events

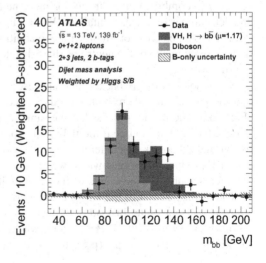

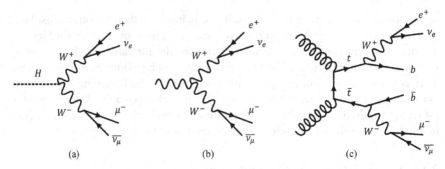

Fig. 8.6 The diagrams of **a** $h \to WW$ **b** a WW-pair production through a Z^0 and **c** a top-quark pair with both W bosons decaying leptonically into $e\mu$

8.1.4 $H \to W^{\pm}W^{\mp*}$

8.1.4.1 Analysis Overview

The branching ratio of the decay channel $h \to W^+W^-$ is about 22%, which is the second largest for $m_h = 125$ GeV. Since the mass of the Higgs boson is less than the sum of two W boson masses (about 161 GeV), one of the two W boson decays virtually ($h \to WW^*$). The analysis using 8 TeV data from the ATLAS collaboration is described in detail in Ref. [6]. The corresponding 13 TeV analysis is given in Ref. [7]; there data analysis procedure is given briefly and refers to the former paper [6]. In this section, some key points of the $H \to WW^*$ analysis are described.

The analysis uses the leptonic decay channel for both of the W bosons ($h \to WW^* \to \ell\nu\ell\nu$) (Fig. 8.6a), where ℓ is either an electron (e) or a muon (μ) in order to reduce background from multijet production $pp \to jets$. Since the multijet final state can be produced with a process with only QCD vertices with strong coupling, the cross section of such production is many orders of magnitude larger than that of $h \to WW^*$ signal. The decay product of the Higgs boson, therefore, is either of $ee, e\mu, \mu\mu$ combinations with two or more neutrinos. The analysis also includes smaller number of events containing $W \to \tau\nu_\tau$ decays where the τ-lepton further decays into an electron or muon with two additional neutrinos. Only the sum of the transverse momenta of the neutrinos (here denoted as $\mathbf{p}_T^{\nu\nu}$) can be measured through missing $\mathbf{p}_T$.

Major sources of the background are resonant-like WW production (Fig. 8.6b) and top-pair events where both top quarks decay leptonically, $t \to Wb, W \to \ell\nu$ (Fig. 8.6c). The former process has the same final state as the signal and is an irreducible background source if the WW pair is produced from a colourless state such as a virtual Z^0 boson. The latter process gives two b-jets and is the main background for VBF production process where we require two jets in the final state and also a significant source for the events with one jet in the final state.

The reconstruction of the Higgs boson mass is not possible because of the neutrinos in the final state. Instead, the transverse mass, m_T, is calculated to estimate the invariant mass of the WW^* system, which uses the transverse components of the kinematic variables: $\mathbf{p}_T^{\nu\nu}(\mathbf{p}_T^{\ell\ell})$, the vector sum of the neutrinos (leptons), and

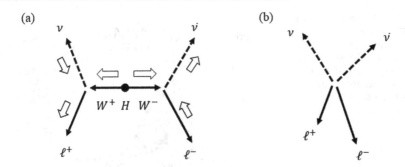

Fig. 8.7 **a** Relation between the spin direction and momentum direction for $H \rightarrow WW^* \rightarrow \ell\nu\ell\nu$ decays. **b** Illustration of typical decay topology in $x - y$ plane (perpendicular to the beam direction)

$E_T^{\ell\ell} = \sqrt{(p_T^{\ell\ell})^2 + (m_{\ell\ell})^2}$. The m_T is defined as

$$m_T = \sqrt{(E_T^{\ell\ell} + p_T^{\ell\ell})^2 - |\mathbf{p}_T^{\ell\ell} + \mathbf{p}_T^{\nu\nu}|^2}.$$

Since $m_T \leq m_h$ and m_h is below twice the W boson mass, $h \rightarrow WW^*$ events will be populated in m_T region below that from resonant WW production (see Fig. 8.8.) The m_T values for top-pair production also tend to be much beyond that from the Higgs decays. This shape difference is used to quantitatively distinguish the signal and background. The peak structure in m_T, for both signal and the WW background, however, is broad. Also, the production rate of WW^* pairs is much smaller than the SM diboson production. Several other features of the signal events are used to reduce background processes.

The number of jets, especially the number of b-jets, is one of such key ingredients to classify event categories. As described in Sect. 8.1.1, at the leading order there is no jet for the ggF processes, while in the VBF processes, each of two incoming quarks emits a vector boson and recoils, giving two jets close to the outgoing beam direction, one for each side. This means that two forward jets are observed, with large separation in rapidity space. Since these forward jets in the VBF processes are jets from light quarks, the background from $t\bar{t}$ production is greatly suppressed by removing events with one or more b-quark jets.

The azimuthal correlation of the two leptons is also used in order to further enhance the signal. Since the Higgs boson is a scalar particle and has no spin, the spin directions of the two W bosons are opposite (Fig. 8.7a). The momentum direction of the charged leptons in $W^- \rightarrow \ell^- \bar{\nu}$ decays tends to be opposite to the spin direction since the anti-neutrino is right-handed and its momentum is aligned to the spin direction. For the W^+, the charged lepton is emitted along the direction of the W^+ spin. As the WW pairs tend to have back-to-back topology in the $x - y$ plane, the direction of the two leptons becomes close as shown in Fig. 8.7b. Also, the invariant mass of the lepton pair, $m_{\ell\ell}$, is peaked around 30–40 GeV while for WW pair production, it is at around 60 GeV, as seen in Fig. 7b in Ref. [6].

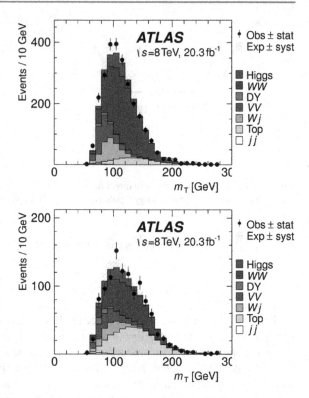

Since the signal-to-background ratio is quite small in WW^* decay channel, the amount of the remaining background is still very large after selecting Higgs-like events using the properties given above. The remaining background depends strongly on the number of accompanied jets. The events are, therefore, classified according to the number of jets: 0-jet, 1-jet, and $\geq$ 2-jet categories. The main background sources for the 0-jet category are irreducible WW production and other diboson production, especially WZ events where one of the leptons is missed. In addition, the events from $W +$ jets production contributes significantly if the jet is misidentified as a lepton. Here, the $W +$ jets process represents higher-order DY events $q\bar{q} \rightarrow W^{\pm}$, i.e., with one or more associated jets. For the 1-jet category, the $t\bar{t}$ production becomes also significant since it produces two b-jets where one of the jets is experimentally not tagged as a b-jet. For the 2-jet events, the major contribution is the $t\bar{t}$ events. The basic idea of how to suppress these background events is described in the next subsection for each category.

8.1.4.2 Background Reduction

- 0-jet category

   After the basic requirement of having two leptons in the final state, significant missing E_T and explicitly requesting no jet, most of the background is the DY process, $pp \rightarrow Z^0/\gamma^* + X$, $Z^0/\gamma^* \rightarrow ee, \mu\mu, \tau\tau$, especially when the two leptons have the same flavour (ee or $\mu\mu$). This background is also significant for $e\mu$ channel, however, since both the τ leptons in $pp \to \tau^+\tau^- X$ processes may decay leptonically, giving a $e\mu$ pair.

In order to further reduce the DY events, the correlation of the lepton pair is used, by requiring p_T of the dilepton system being high: $p_T^{\ell\ell} > 30\,\text{GeV}$ (Fig. 7a in Ref. [6]). Since the Z^0/γ^* in the DY processes are produced from $q\bar{q}$ annihilation, each of the quarks coming from the incoming protons, the transverse momentum of the produced Z^0/γ^* tend to be small and the lepton pair from the decay tends to be produced back-to-back in the $x - y$ plane.

The missing E_T may arise from background processes through mismeasurement of the energy or momentum of the final state particles, i.e. two leptons. In such cases the missing E_T tends to be aligned to the momentum direction of these particles. A few requirements are applied based on the relative momentum of the missing $\mathbf{p}_T$ to the leptons.

Finally, the azimuthal correlation requirement ($\phi_{\ell\ell} < 1.8$) and the mass of the dilepton system $m_{\ell\ell} < 55\,\text{GeV}$ are required to select events with $H \to WW^*$ topology as described above.

- **1-jet category**

  The event selection for the 1-jet category is very similar to that for the 0-jet events apart from a few points: the required jet should not be tagged as a b-jet; $\mathbf{p}_T^{\ell\ell}$ is replaced to $\mathbf{p}_T^{\ell\ell j}$, adding the momentum of the jet; and additional requirement on the $m_{\tau\tau}$ variable is imposed: $m_{\tau\tau} < m_Z - 25\,\text{GeV}$ where m_Z is the mass of the Z^0 boson. The $m_{\tau\tau}$ variable is calculated by using so-called "collinear approximation" assuming that the leptons are from the decay of τ leptons originated from Z^0 and the momentum of the rest of the τ decay products, two neutrinos for each decay, are estimated by projecting the missing p_T vector to the two lepton directions.

- **2-jet category**

  The signal-to-noise ratio for two-jet VBF categories is much smaller than the other categories at the stage after dilepton + missing E_T selection. In order to enrich the signal, a machine-learning technique (boosted decision tree, BDT) is used. The detail of the technique is beyond the scope of this book. Here we merely explain the main variables used as inputs for the machinery. Two variables related to the forward-going two jets, the jet-jet mass m_{jj} and the rapidity difference between the two jets y_{jj}, play main role in the selection since the two jets in the VBF process tend to have large values. Some other variables related to the angular order of the VBF jets and the decay products of the Higgs boson are used to enrich the VBF process, based on the fact that the Higgs boson is produced in between the two jets, each of which goes into near the outgoing beam direction on the opposite sides (see Fig. 8.2b). In addition, since the VBF is a quark induced process without QCD vertex (see Sect. 8.1.1), the amount of the initial and final state radiations from partons are largely suppressed with respect to the main background process, the $t\bar{t}$ production. The vector sum of $\mathbf{p}_T$ over hard objects in an event is sensitive to the amount of such radiation since the size of such vector indicates the amount of recoil received by the objects.

Figure 8.8 shows the m_T distribution of the events after all the selection for $e\mu$ channel. A clear excess over the sum of the background is observed for both 0- and 1-jet categories. The amount of the excess divided by the expected number of events predicted by the Standard Model Higgs boson production cross section is called signal strength parameter (denoted as μ). The value of μ is extracted from the fit to m_T distributions of all the event categories after fixing the background distributions including their normalisations, as described below.

8.1.4.3 Background Estimation

It is difficult to determine the amount of background events through template fit assuming the shape of the signal and background and determining the normalisation of each contribution through the fit, since the m_T distribution for the signal is relatively broad and the shape is somewhat similar for the signal and some of the background events as seen in Fig. 8.8. The background contribution is, therefore, estimated by using event distributions in control regions where some of the selection criteria are inverted so that there is no overlap in events between the signal and control regions.

In this analysis, the control regions are prepared for each process for each category of events (0, 1, or 2 jets, $e\mu$ or $ee + \mu\mu$ final states) and for each background process (WW, top, Drell-Yan, etc.). Instead of going through all of them, we pick up a few most relevant ones.

For example, the normalisation of the WW contribution is obtained by events in high $m_{\ell\ell}$ region, $55 < m_{\ell\ell} < 110$ GeV for the 0-jet category so that the purity of the WW contribution is improved, while keeping similar event selection criteria to the signal region. The remaining background sources from non-WW processes in this control region are subtracted by using simulated events.

The strongest constraint for normalising $t\bar{t}$ contribution comes from 1-jet category $e\mu$ final state, but requesting one b-tagged jet explicitly, since all top quarks practically decay to the bW final state. In addition, the requirement on lepton is tightened by requesting $m_T^\ell > 50$ GeV, where m_T^ℓ is defined as the mass between one of the leptons and missing p_T vector on the $x - y$ plane. It is meant for reconstructing the transverse mass of the W bosons from the top-quark decays. After applying these criteria, the control region consists almost fully of top-quark production. Thus determined background fraction gives consistent results with simulation for most of the control regions, despite the fact that the event selection for $H \rightarrow WW$ may be at the corner of the phase space for the background processes.

After repeating similar exercises for other event categories, the normalisation factors for the background processes as well the signal contribution are finally fixed by performing a simultaneous likelihood fit, where some of the normalisation factors are allowed to shift while others are fixed. The final result for the 8 TeV analysis gives $\mu = 1.09^{+0.16}_{-0.15}(\text{stat})^{+0.17}_{-0.14}(\text{syst})$. The main sources of the systematic uncertainties are theoretical origin, like the cross section prediction of the signal itself, since the strength parameter is the cross section ratios of measurement to prediction. For the 13 TeV analysis, the statistical uncertainty was improved and became lower than the total systematic uncertainties.

8.2   Search for Physics Beyond the Standard Model

One of the main goals of the high-energy experiments is the discovery of new phenomena so that there is plenty of data analysis for physics beyond the standard model (BSM). From among them, we explain SUSY (supersymmetry) and resonance searches, which are typical BSM searches; their idea can be applicable to data analysis for other BSM.

8.2.1   SUSY

Many searches for phenomena beyond the Standard Model target a signal that does not make any resonance of new particles. One of the best examples is SUSY search. The supersymmetry is a new fermion-boson symmetry, where new fermion (boson) partners are introduced for all standard model bosons (fermions). The supersymmetric partners of electron (e), weak boson (W), quark (q), and gluon (g) as examples are scalar electron (selectron, $\tilde{e}$), wino ($\tilde{W}$), scalar quark (squark, $\tilde{q}$), and gluino ($\tilde{g}$), respectively.[6] They have the same mass as their partners; however, we have not seen such particles so far. This symmetry is assumed to be broken and the mass of supersymmetric partners can be heavy. The lightest supersymmetric particle (LSP) is assumed to be neutral and stable (under R-parity conservation) and cannot be detected so that the LSP is a good candidate for dark matter. This is one of the motivations for SUSY models.

In R-parity conversed SUSY models, a pair of SUSY particles, which are new particles for us so far, can be produced in the LHC and then each SUSY particle decays eventually in SM particles and one LSP. Due to the existence of the LSP in the decay chain, we cannot reconstruct the mass of any SUSY particles which are produced in the decay chain. Even in such cases, there are several useful variables to search for the SUSY signal and more variables are being developed.

Since the LHC is a pp collider, we expect large production cross sections of SUSY signal via the strong interaction: $gg \to \tilde{g}\tilde{g}$, $gg \to \tilde{q}\tilde{q}$, and $gq \to \tilde{g}\tilde{q}$ as shown in Fig. 8.9. The search for SUSY with these channels is of importance in the LHC. We focus on the search for SUSY through the $gg \to \tilde{g}\tilde{g}$ production process, where we assume that the other SUSY particles except for the lightest neutralino $\tilde{\chi}_1^{\,0}$ are heavier than gluino $\tilde{g}$.

In such a simple scenario, gluinos decay into two quarks plus $\tilde{\chi}_1^{\,0}$ via $\tilde{g} \to q\tilde{q}^* \to qq\tilde{\chi}_1^{\,0}$, giving four quarks (including anti-quarks) and two $\tilde{\chi}_1^{\,0}$ in the final state. In this analysis, we require four or more high p_T jets and a large missing transverse energy. Additional high p_T jets might come from the initial and final state radiations. In nominal SUSY searches, there are two useful variables to separate signal events from background events: missing transverse energy E_T^{miss} and so-called "effective" mass

---

[6] The naming convention of supersymmetric partners is the prefix of "s-" for fermions and the postfix of "-ino" for bosons.

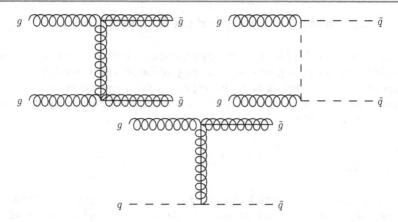

Fig. 8.9 Feynman diagrams of SUSY production via strong interaction in the pp collider

m_{eff}. The m_{eff} variable is defined to be the scalar sum of the transverse momentum of jets and $E_{\text{T}}^{\text{miss}}$: $m_{\text{eff}} = \sum_{\text{jet}} p_{\text{T}} + E_{\text{T}}^{\text{miss}}$. The number of jets that are added in the summation depends on analyses, for example, up to four in the p_{T} order. The m_{eff} variable corresponds to the mass of the SUSY particle pair initially produced. Figure 8.10 shows the m_{eff} distribution for gluinos and squarks search in ATLAS [8]. SUSY signal events can have dumps in the high $E_{\text{T}}^{\text{miss}}$ and m_{eff} regions. This is a typical SUSY signal for which we have searched.

In practice, we are moving to a more complex analysis to improve the signal sensitivity since any SUSY signal has not been seen in the LHC. We have adopted multivariate analysis techniques like BDT, deep learning (DL), etc. The variables of $E_{\text{T}}^{\text{miss}}$ and m_{eff} are one of the input variables to them. In these analysis techniques, not only each input variable but also the correlation of input variables are utilised to separate the signal from the background. Since the selection criteria are determined by using the MC events, for example, BDT or DL is trained with MC samples, we are careful that the correlation of variables in MC events should be similar to that in the real data as much as possible. Such checks are required to adopt the multivariate analysis technique.

8.2.2   Resonance Search

As the charmonium was discovered by a resonance of a pair of electrons and the Higgs boson was recently discovered by peaks of a pair of γs and 4 leptons, it is historically evident that looking for any resonances of the new particles is the one of the most effective and the easiest ways to search for new physics independent of the theoretical models.

The distribution of the invariant mass for oppositely charged muon pairs with transverse momentum above 4 GeV and pseudorapidity $|\eta| < 2.5$ and selected by muon triggers at the ATLAS experiment is shown in Fig. 8.11. In case two reconstructed muons are originated from a particle with narrow decay width such as J/ψ,

Fig. 8.10 Distribution of m_{eff} for a 6-jet region in the SUSY search. Reprinted from [8] under the Creative Commons Attribution 4.0 International License © 2018 CERN, for the ATLAS Collaboration. The points with bars show observed data. The histograms show the MC background predictions prior to the fits. The arrows indicate the values at which the requirements on m_{eff} are applied. The expected distribution for a SUSY signal model point is shown with a dotted-line (masses in GeV)

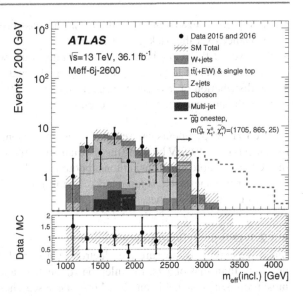

Fig. 8.11 Dimuon invariant mass distribution for oppositely charged muon pairs with transverse momentum above 4 GeV and pseudorapidity $|\eta| < 2.5$ and selected by muon triggers. Reprinted under the Creative Commons Attribution 4.0 International License from [9] © 2015 CERN for the benefit of the ATLAS Collaboration. The resonances of J/ψ, ψ', Υ resonances, and Z are clearly visible in this distribution

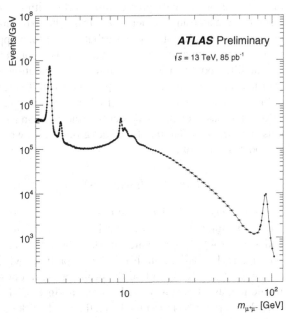

ψ', Υ, and Z boson, the invariant mass reconstructed by momenta and energies of two muons are measured to be around the mass of the resonance particle. On the other hand, the invariant mass reconstructed by candidates of two muons (including charged particles faking as muons) which are not originated from a decay of particle distributes continuously according to the combination of values of momenta and energies of the two muons. From this example of the search for the peak of "known" particle, one can learn that

- more precise measurement of the invariant mass provides a sharper peak of the resonance over the backgrounds,
- the level of the reducible backgrounds due to the wrong measurement such as fakes needs to be lowered as much as possible, and
- the distribution of irreducible backgrounds needs to be under-controlled to estimate the number of background events.

The LHC experiments can search for the new resonances predicted by the BSM hypothesis up to 10 TeV by using the invariant mass reconstructed by the combination of the two or more electrons, muons, photons, and jets, which includes the decays of heavier particles such as top quarks. The following sections show two examples of BSM resonance searches.

8.2.2.1 Dilepton Resonances

Since we expect to measure the electron energy and the muon momentum more precisely than that of jets, the dilepton (dielectron and dimuon) final state is the most promising channel in any BSM resonance searches. From the theoretical point of view, various models predict resonances with decay into dileptons and can be categorised according to their spin. Thus, the experimentalists first search for any excesses in the dilepton mass distribution and then apply the result of the searches to the interpretation of models with such new resonances.

The filled points in Fig. 8.12 show the distribution of the dielectron and dimuon invariant mass ($m_{\ell\ell}$) for events passing the full selection using 139 fb^{-1} of pp collision data collected at $\sqrt{s} = 13$ TeV with the ATLAS detector [10]. The event selection is based on the quality cuts of the electron and muon, their p_T, and fiducial cuts. The $m_{\ell\ell}$ distribution of the backgrounds, shown as red solid lines in Fig. 8.12, is modelled by formula of

$$f(m_{\ell\ell}) = f_{\mathrm{BW},Z}(m_{\ell\ell}) \cdot (1 - x^c)^b \cdot x^{\sum_{i=0}^{3} p_i \log(x)^i}, \tag{8.1}$$

where $x = m_{\ell\ell}/\sqrt{s}$ and b, c, and p_i with $i = 0, ...3$ are the parameters determined by the fit. The function $f_{\mathrm{BW},Z}(m_{\ell\ell})$ is Breit-Wigner function with $m_Z = 91.1876$ GeV and $\Gamma_Z = 2.4952$ GeV, which models the line shape of the resonance of the Z boson at high mass region. If new heavy particles with pole masses of 1.34, 2, and 3 TeV existed, one could find the peaks of the dilepton mass over the background prediction, as shown as dashed curves in Fig. 8.12. In the prediction of these new particles, zero width is assumed, i.e., the width of the distributions is only due to the detector resolutions. Since the electron energy measured from the electromagnetic shower is more precise than the momentum measurement for a charged particle in the energy region of our interests, dielectron mass reconstructed from the energy measurement has better resolution than that from the momentum measurement. For the dimuon channel, on the other hand, only the momentum measurement is available. Therefore, the mass resolution of dielectron is better than that of dimuon. Figure 8.12 does not show any sign of a signal from the new particle. If you want to quantify if a signal exists or not, you can calculate the probability that the data are compatible

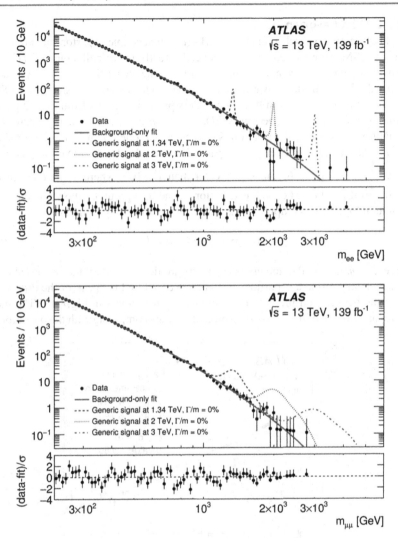

Fig. 8.12 Distribution of the **a** dielectron and **b** dimuon invariant mass for events passing the full selection. Reprinted under the Creative Commons Attribution 4.0 International License from [10] © 2019 The Author. Generic zero-width signal shapes, scaled to 20 times the value of the corresponding expected upper limit at 95% CL on the fiducial cross section times branching ratio, with pole masses of $m_X = 1.34$, 2, and 3 TeV, as well as background-only fits, are superimposed. The data points are plotted at the centre of each bin. The error bars indicate statistical uncertainties only. The differences between the data and the fit results in units of standard deviations of the statistical uncertainty are shown in the bottom panels

with the background-only hypothesis as is described for $H \rightarrow \gamma\gamma$ peak search (see Sect. 8.1.2).

8.2.2.2 Dijet Resonances

New heavy particles, such as excited quarks (q^*), that couple to partons are predicted in many BSM theories and can be produced directly in pp collisions at LHC and decayed into partons. Events of this kind of a new heavy particle produce a peak in the distribution of the dijet invariant mass (m_{jj}). On the other hand, since, in the SM, the production of jet pairs in hadron colliders primarily results from $2 \rightarrow 2$ parton scattering processes described by QCD, a smooth and monotonically decreasing distribution for the m_{jj} distribution is expected. The filled points in Fig. 8.13 show the m_{jj} distribution for events with $p_T > 150$ GeV for the two leading jets, with $|y^*| \equiv \frac{1}{2}|y_1 - y_2| < 0.6$, and m_{jj} greater than 1.1 TeV, where y_1 and y_2 are the rapidity of dijet [11]. The m_{jj} distribution of the backgrounds, shown as the solid red line in Fig. 8.13, is empirically known to be predicted by formula:

$$f(x) = p_1(1 - x)^{p_2} x^{p_3 + p_4 \ln x}, \tag{8.2}$$

where $x = m_{jj}/\sqrt{s}$. Parameters of p_1 to p_4 are determined by the fit to real data. If a new heavy resonance particle existed, one could find the peak of the dijet mass above the background prediction, as shown as open points in Fig. 8.13. The most discrepant interval of the m_{jj} distribution of the data comparing with the background

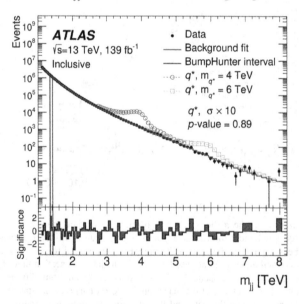

Fig. 8.13 The reconstructed dijet mass distribution, m_{jj}, is shown for events with $p_T > 150$ GeV for the two leading jets, with $|y^*| < 1.2$, and m_{jj} greater than 1.1 TeV (filled points). Reprinted under the Creative Commons Attribution 4.0 International License from [11] © CERN, for the benefit of the ATLAS Collaboration. The solid line depicts the background prediction from the sliding-window fit. The vertical lines indicate the most discrepant interval, for which the p-value is 0.89 as reported in the figure. The expected contributions for q^* signal with a mass of 4 and 6 TeV are overlaid, normalised to 10 times their predicted cross section. The lower panel shows the bin-by-bin significance of the data-fit discrepancy, based only on statistical uncertainties

prediction is indicated by the two vertical blue lines in Fig. 8.13. The p-value for the most discrepant interval is calculated to be 0.89.

References

1. LHC Higgs Cross Section Working Group, Heinemeyer, S., Mariotti, C., Passarino, G., Tanaka, R.(eds.): CERN-2013-004. arXiv:1307.1347 [hep-ph]
2. Aad, G., et al.: ATLAS collaboration. Phys. Lett. B **716**, 1 (2012). https://doi.org/10.1016/j.physletb.2012.08.020. arXiv:1207.7214 [hep-ex]
3. Chatrchyan, S., et al.: CMS collaboration. Phys. Lett. B **716**, 30 (2012). https://doi.org/10.1016/j.physletb.2012.08.021. arXiv:1207.7235 [hep-ex]
4. Aad, G., et al.: ATLAS collaboration. Phys. Lett. B **726**, 88 (2013). Erratum: Phys. Lett. B **734**, 406 (2014). https://doi.org/10.1016/j.physletb.2014.05.011, https://doi.org/10.1016/j.physletb.2013.08.010. arXiv:1307.1427 [hep-ex], Auxiliary materials: https://atlas.web.cern.ch/Atlas/GROUPS/PHYSICS/PAPERS/HIGG-2013-02
5. Aad, G., et al.: ATLAS collaboration. Eur. Phys. J. C **81**(2), 178 (2021). https://doi.org/10.1140/epjc/s10052-020-08677-2. arXiv:2007.02873 [hep-ex]
6. Aad, G., et al.: ATLAS collaboration. Phys. Rev. D **92**(1), 012006 (2015). https://doi.org/10.1103/PhysRevD.92.012006. arXiv:1412.2641 [hep-ex]
7. Aaboud, M., et al.: ATLAS collaboration. Phys. Lett. B **789**, 508–529 (2019). https://doi.org/10.1016/j.physletb.2018.11.064. arXiv:1808.09054 [hep-ex]
8. Aaboud, M., et al.: ATLAS collaboration. Phys. Rev. D **97**(11), 112001 (2018). https://doi.org/10.1103/PhysRevD.97.112001. arXiv:1712.02332 [hep-ex]
9. ATLAS collaboration, ATL-PHYS-PUB-2015-037
10. Aad, G., et al.: ATLAS collaboration. Phys. Lett. B **796**, 68–87 (2019). https://doi.org/10.1016/j.physletb.2019.07.016. arXiv:1903.06248 [hep-ex]
11. Aad, G., et al.: ATLAS collaboration. JHEP **03**, 145 (2020). https://doi.org/10.1007/JHEP03(2020)145. arXiv:1910.08447 [hep-ex]

Statistics

A

The detail of the calculation to obtain the mean, variance, etc., which is not shown in the main text, is given in this section. This is useful for the beginners in statistics.

A.1 Binomial Distribution

$$\mu = \sum_n n P(n)$$

$$= \sum_n n \cdot \frac{N!}{n!(N-n)!} p^n (1-p)^{N-n}$$

$$= \sum_n n \cdot \frac{N}{n} \cdot \frac{(N-1)!}{(n-1)!(N-1-(n-1))!} \cdot p \cdot p^{n-1}(1-p)^{N-1-(n-1)}$$

$$= Np \cdot \sum_n \frac{(N-1)!}{(n-1)!(N-1-(n-1))!} \cdot p^{n-1}(1-p)^{N-1-(n-1)}$$

$$= Np \cdot (p+1-p)^{N-1} = NP \tag{A.1}$$

$$\sigma^2 = \sum_n (n-\mu)^2 P(n)$$

$$= \sum_n n^2 P(n) - \sum_n \mu^2 P(n)$$

$$= \sum_n n^2 \cdot \frac{N}{n} \cdot \frac{(N-1)!}{(n-1)!(N-1-(n-1))!} \cdot p \cdot p^{n-1}(1-p)^{N-1-(n-1)} - (Np)^2$$

$$= Np \sum_n ((n-1)+1) \cdot \frac{(N-1)!}{(n-1)!(N-1-(n-1))!} p^{n-1}(1-p)^{N-1-(n-1)} - (Np)^2$$

$$= Np((N-1)p+1) - (Np)^2$$

$$= Np(1-p) \tag{A.2}$$

© The Editor(s) (if applicable) and The Author(s) 2022
K. Hanagaki et al., *Experimental Techniques in Modern High-Energy Physics*,
Lecture Notes in Physics 1001, https://doi.org/10.1007/978-4-431-56931-2_A

A.2   Poisson's Distribution

$$\mu = \sum_{n}^{\infty} n \cdot \frac{\mu^n e^{-\mu}}{n!}$$

$$= \mu e^{-\mu} \sum_{n}^{\infty} \frac{\mu^{n-1}}{(n-1)!}$$

$$= \mu \tag{A.3}$$

$$\sigma^2 = \sum_{n}^{\infty} (n-\mu)^2 \cdot \frac{\mu^n e^{-\mu}}{n!}$$

$$= \sum_{n}^{\infty} n^2 \cdot \frac{\mu^n e^{-\mu}}{n!} - \mu^2$$

$$= \sum_{n}^{\infty} \mu e^{-\mu} \cdot n \cdot \frac{\mu^{n-1}}{(n-1)!} - \mu^2$$

$$= \mu e^{-\mu} \sum_{n}^{\infty} (n+1) \cdot \frac{\mu^n}{n!} - \mu^2$$

$$= \mu^2 + \mu - \mu^2$$

$$= \mu \tag{A.4}$$

A.3   Maximum Likelihood Method

$$E(\hat{\mu}) = \int \cdots \int \hat{\mu} \times \prod_{i=1}^{n} \frac{1}{\sqrt{2\pi}\sigma} \exp\left(-\frac{(x_i-\mu)^2}{2\sigma^2}\right) dx_1 \cdots dx_n$$

$$= \int \cdots \int \left(\frac{1}{n}\sum_{i=1}^{n} x_i\right) \prod_{i=1}^{n} \frac{1}{\sqrt{2\pi}\sigma} \exp\left(-\frac{(x_i-\mu)^2}{2\sigma^2}\right) dx_1 \cdots dx_n$$

$$= \frac{1}{n}\sum_{i=1}^{n} \left(\int \frac{x_i}{\sqrt{2\pi}\sigma} \exp\left(-\frac{(x_i-\mu)^2}{2\sigma^2}\right) dx_i \prod_{j\neq i} \int \frac{1}{\sqrt{2\pi}\sigma} \exp\left(-\frac{(x_j-\mu)^2}{2\sigma^2}\right) dx_j\right)$$

$$= \frac{1}{n}\sum_{i=1}^{n} \mu$$

$$= \mu \tag{A.5}$$

Printed in the United States
by Baker & Taylor Publisher Services